T0092045

People of the Screen

People of the Screen

How Evangelicals Created the Digital Bible and How It Shapes Their Reading of Scripture

JOHN DYER

OXFORD
UNIVERSITY PRESS

Oxford University Press is a department of the University of Oxford. It furthers the University's objective of excellence in research, scholarship, and education by publishing worldwide. Oxford is a registered trade mark of Oxford University Press in the UK and certain other countries.

Published in the United States of America by Oxford University Press
198 Madison Avenue, New York, NY 10016, United States of America.

Library of Congress Cataloging-in-Publication Data
Names: Dyer, John, 1979- author.
Title: People of the screen : how evangelicals created the digital
Bible and how it shapes their reading of scripture / John Dyer.
Description: New York, NY, United States of America : Oxford University Press, [2023] |
Includes bibliographical references and index.
Identifiers: LCCN 2022027323 (print) | LCCN 2022027324 (ebook) |
ISBN 9780197636350 (hb) | ISBN 9780197636374 (epub)
Subjects: LCSH: Bible—Criticism, interpretation, etc.—Data processing. |
Bible and technology. | Evangelicalism. | Bible—Reading.
Classification: LCC BS534.8 .D94 2023 (print) | LCC BS534.8 (ebook) |
DDC 220.6—dc23/eng/20220810
LC record available at https://lccn.loc.gov/2022027323
LC ebook record available at https://lccn.loc.gov/2022027324

DOI: 10.1093/oso/9780197636350.001.0001

1 3 5 7 9 8 6 4 2

Printed by Sheridan Books, Inc., United States of America

Contents

Preface, in Which I Admit to Being an Evangelical Bible Programmer Studying Evangelical Bible Programmers vii
Acknowledgments ix

1. Introduction 1
   - A New Kind of Technology Company 1
   - Bible Tech Revolutions 1
   - Techno-Evangelicals 3
   - The State of the Digital Bible 7
2. Studying Technology and Faith 12
   - Early Approaches to Technology Culture 12
   - Social Shaping of Technology 13
   - Digital Religion 19
   - Study Groups and Approach 24
3. Evangelicals, the Bible, and Technology 28
   - Exploring Evangelicalism 29
   - The Supremacy of the Bible 36
   - Hopeful Attitude toward Technology 39
   - Entrepreneurialism in Business and Ministry 44
   - Pragmatism in Mission and Outcomes 48
   - Missions Movement 50
   - Political Engagement 52
   - Biblically Focused, Hopeful Entrepreneurial Pragmatism 56
4. Four Waves of Bible Software Development 58
   - The Preconsumer Academic Era, 1950s–1970s 59
   - The Desktop Pastoral Era, 1980 Onward 61
   - The Popular Internet Era, 1995 Onward 70
   - The Mobile App Era, 2000s Onward 77
   - Summary and Significance 83
5. Programmers and the Business of Bible Software 86
   - People: How Do Evangelicals Become Bible Software Developers? 87
   - A Range of Views on Personal Technology Usage 92
   - Product: What Is the Goal of Bible Software? 94
   - Process: How Is Bible Software Made? 104

Profit: How Does Bible Software Sustain Itself? 116
The Digital Bible as Evangelical Mission 122

6. A Portrait of Evangelical Bible Readers 124
Focus Group Approach and Summary 125
Social Context: Relationships and the Bible 126
Forms of Bible Engagement by Media 131
Software: Replicating, Upgrading, and Enabling New Forms of Bible Engagement 135
Hardware: The Convenience of Mobile Devices 151
The Fluid Usage of Print and Digital 157

7. The Influence of Digital on Evangelical Reader Behavior 160
The State of Reading on Screen 161
Print versus Screen: Form and Features 161
Comprehension Issues, but Only for Some 164
From Comprehension to Hermeneutics and Experience 166
Technologically Enhanced Daily Bible Reading 168
Focusing on the Bible in a Digital World 172
The Social Shaping of Bible Software's Impact 179

8. Conclusions 181
The Digital Bible in Modern Faith 186

Appendix: Bible Software List 189
Desktop Software, 1982–1987 189
Desktop Software, 1987–2000 190
Mobile Era 192
Notes 195
Works Cited 231
Index 257

Preface, in Which I Admit to Being an Evangelical Bible Programmer Studying Evangelical Bible Programmers

It seemed almost inevitable.

I grew up watching my dad tinker with computers and my mom read the Bible. Over time, as I explored these inheritances they seemed to overlap and converge. The Bible looked like a database with vast amounts of information waiting to be organized, and technology seemed able to solve the world's deepest problems. Something in me said they belonged together.

Finally, one day, I gave in and wrote my first Bible app.

Sociologists and anthologists have a term—reflexivity—to describe the interconnected relationship between a researcher and the object of her research. No researcher can be entirely objective, but the more one admits to the biases and perspectives one has, the better chance one has at not allowing that perspective to unduly, or unhelpfully, misdirect one's research. The next few paragraphs are my attempt to foreground some of my experiences and orientation.

My religious upbringing has primarily taken place within the broad and somewhat amorphous culture of American evangelicalism. During my youth, my parents took our family to several churches within the Baptist and Evangelical Free (which, incidentally, does not mean "free of evangelicals") denominations along with several nondenominational churches with words like "Bible," "Fellowship," and "Community" in their names. In addition, I briefly served as a youth pastor for a nondenominational Bible church, which led me to attend an evangelical seminary for my master's work. I now work at an evangelical seminary overseeing communications, enrollment, and online education, as well as teaching in the areas of theology, technology, and sociology. Because of the painful associations of white evangelicalism with some of our nation's worst sins, there have been times where I might want to, as Mark Noll has written, "demit my status as an evangelical if only I knew where to send in my resignation." And yet, over time, I have become more comfortable with my evangelical identity both as a complex part of my heritage and as a helpful starting point for the research in this book.

Beyond my own religious upbringing and its emphasis on the importance of the Bible, it is also important to note that software development has been my profession for nearly two decades, during which I have helped develop tools used by companies like Apple, Facebook, Twitter, and Anheuser-Busch. As mentioned, I have also created Bible software both for myself as a mode of personal inquiry and as compensated work. For example, I created biblewebapp.com, which offers similar functionality to desktop Bible software and yallversion.com (a play on youversion.com), which highlights all the second-person plurals in Greek and Hebrew and replaces them with regional variants such as "y'all" (southern United States), "yinz" (Pittsburgh, PA), and "you lot."

The relevance of these experiences to this book is I am uniquely situated in the part of a Venn diagram where "evangelical" and "programmer" overlap. This allows me to speak both the language of evangelicalism and the language of programming and has given me unique access to Bible software companies and their personnel. In one sense, I share many of the same convictions and mission of the people and companies I explore in this book. At the same time, I am concerned that parts of evangelicalism have gone deeply astray from its core convictions. My position then allows me to directly explore an underresearched aspect of American religion—the role of evangelicals in Bible software—while also using that to reflect on the nature of evangelicalism, both as a system of beliefs and as a reaction to cultural change, in this case, the rapid development of technology.

Acknowledgments

This book, which grew out my doctoral work at Durham University, would not have been possible without the insight, kindness, and encouragement of my advisors, Mathew Guest and Pete Phillips. From Durham to San Antonio to Boston, in person and online, they skillfully guided me, showing me how to be a careful and engaged scholar. Dr. Guest's intimate familiarity with sociology of religion and Dr. Phillips's enthusiasm for digital religion made this a richer and more enjoyable project.

Several other scholars have offered their help along the way, including Heidi Campbell, Tim Hutchings, Jeff Siker, Ruth Perrin, Samuel Perry, and the entire team at CODEC. I am also grateful to the Bible software companies and their leaders, Bob and Dan Pritchett at Faithlife, Rachel Barach and Stephen Smith at Bible Gateway, and Terry Storch at YouVersion, who generously offered their time and access to their teams to make this study possible. In addition, the churches, pseudonymously known as City Bible Church, Petra Community Church, and Hidden Baptist Church, also deserve thanks for giving me class time in which to study their congregants.

I have also benefited enormously from the support of my employer, Dallas Theological Seminary, especially my supervisor Mark Yarbrough, who gave me time to complete the initial doctoral thesis version of this work and its adaptation into a book form.

Friends within my faith community have also been supportive of me while I was writing. I think particularly of my long-time friends Dave Furman and Frank Barnett, who have offered decades of wisdom and good humor, and Rhett Smith and Scott McClellan whose online friendship thankfully made the transition to the pub. We are also grateful to Andy and Miranda Byers for making our transition to Durham a delightful adventure and to Nick and Tina Bucknall for showing us the fullness of British hospitality.

I am deeply appreciative of Oxford University Press for choosing to pick up this title and for Theodore Calderara's deft editorial guidance in getting the text down to a size that we hope is useful for scholars and students.

Finally, the family I have made with the recently minted Dr. Amber Dyer, and our two children Benjamin and Rebecca Dyer, is the joy and highlight of my life. I am grateful for the patience my children have offered while I write and for their incredible indexing skills when it was complete. Their curiosity, wit, and spark have made me a better thinker and person. To my wife Amber, thank you for being a brilliant scholar in your field, a lovely mother and wife in our home, a fearless leader wherever you go, and an irreplaceable friend in my life.

1
Introduction

A New Kind of Technology Company

In the summer of 1979, just a few years after Jimmy Carter brought the term "born-again" into the mainstream American lexicon and Steve Jobs made the home computer a part of everyday life, two engineers at Intel hatched a plan to create a new kind of technology company.[1] Kent Ochel and Bert Brown's new endeavor would combine their religious faith and their lifelong desire to build their own company, enabling them to do something unprecedented—they would bring the Bible into the digital age and put it on every personal computer in the world. Early the next year in Austin, Texas, at the crossroads of the American Bible Belt and the burgeoning computer industry, they created Bible Research Systems and set to work merging their technical know-how with their love of scripture.

In January 1982, they released the first version of The Word Processor for the Apple IIe, making it "the first commercial Bible study software on the market."[2] *Softalk Mag* hailed it for including a complete and searchable text of the King James Bible, promising it would "aid the serious Bible student" and comparing Ochel and Brown's accomplishments to Gutenberg's printing press.[3] As the personal computer industry and Bible software market grew alongside one another in the 1980s, scholars and religious people alike began to wonder if computers might fundamentally change religion, and more specifically, how the shift from printed books to electronic media would transform the practices of Christians, who for centuries had been called "the people of the book."[4]

What will happen to Christians as they become "the people of the screen"?

Bible Tech Revolutions

Before the advent of Bible software, Christianity had undergone two previous shifts in the use of media. The first was the shift from the scroll to the codex

People of the Screen. John Dyer, Oxford University Press. © Oxford University Press 2023.
DOI: 10.1093/oso/9780197636350.003.0001

in the first century, which some scholars argue became an identity marker for early Christians that distinguished them from Jews and pagans. Although the codex was not initially considered worthy of something as weighty as Holy Scripture, early Christians appear to have found the codex easier to use: faster for finding passages, capable of holding more information, and better suited for travel.[5] On top of the technology of the codex, Christians added various innovations such as visual flourishes, parallel columns, notes, and other study tools, including divisions that are similar to our modern chapters. But the expense and time it took to produce a full copy of the Bible meant that average Christians could not afford their own Bibles and only clergy had the privilege of reading from them.

The second major technology change came in the shift from the handwritten codex to the printing press in the fifteen century. Historian Elizabeth Eisenstein argued that the printing press was instrumental in spurring many of the large-scale cultural changes in Europe, including the Scientific Revolution and the Protestant Reformation.[6] Indeed, Luther, who wrote his Ninety-Five Theses (1517) more than seventy-five years after the construction of Gutenberg's first printing press (1440), himself declared the technology to be "God's highest and extremest act of grace, whereby the business of the Gospel is driven forward." As with the codex, printers took advantage of new technology to add new features to the Bible that would go on to shape how the Bible would be read. Perhaps the most powerful was the creation and standardization of the chapter-and-verse numbering system. Today, these numbers are so common that the average reader might assume they were part of the original biblical writings, but the modern versification scheme we use today was created by French scholar and printer Robert Estienne for his 1551 edition of the Greek New Testament.[7] His system quickly spread to other versions, including the English Geneva Bible in 1560, and the chapter:verse system is now an essential part of everything from biblical scholarship to sermon notes to Christian merchandising. The printing press also drastically reduced the cost of Bibles, enabling the creation of more translations, study features, and other innovations.

The electronic or digital revolution represents a third shift that is just getting underway. Religion professor Bryan Bibb writes that "the current shift from *codex* to *screen* will be every bit as decisive as the historic shift from *scroll* to *codex* in the Greco-Roman world, or the shift from hand-lettered to printed manuscripts in the Late Middle Ages."[8] If computers and the internet are truly enabling a similarly spectacular social change in the present, will the shift from the printed Bible to the digital Bible cause Christianity to undergo another Reformation of sorts? What innovations might be built on top of the screen that will reshape the way Christians read their sacred text?

Techno-Evangelicals

The digital revolution is too new to truly assess whether it will create a new Reformation, but we are far enough into the digital Bible's creation and use to begin examining the actors behind industry. Technological change is sometimes framed in terms of the technology itself, with authors asking questions like "Is Google Making Us Stupid?"[9] But instead of attributing personhood and agency to technology itself, we can assess technology more clearly by pulling back the veil and investigating the people who create Bible software—programmers, entrepreneurs, and Christian business leaders—and their role in reshaping how modern readers encounter scripture. Who are the Gutenbergs of today, we might ask, that are bringing the Bible to us on screen, and what are their motivations and beliefs? How do they react to changes in the overall technology market and the requests of their customers?[10]

What we will find is that of all Christian traditions, evangelicals stand apart for their involvement in both the production and consumption of digital Bibles. In fact, as this book will show, after the initial wave of Bible programming experiments in the 1950s and 1960s, nearly all the major companies and ministries involved in the creation of digital Bibles—beginning with academic software in the 1970s and continuing through the personal computing era of the 1980s, the launch of the internet in the 1990s, and the mobile era of this century—have roots in evangelicalism. In addition, the churches and individuals most likely to incorporate digital Bibles in their faith are evangelicals. Muslims, Hindus, and other religious groups create applications for their followers, and nonevangelical Christians have created Bible software, but the most commercially successful desktop applications, the most highly trafficked websites, and the highest ranked mobile Bible apps were all created by evangelical individuals, companies, or ministry organizations. These technological entrepreneurs brought to the digital Bible enterprise a distinctly evangelical outlook on the Bible as an object and as a religious text, and their beliefs about how culture, media, and religion interact were mutually shaped by the move into digital media.

In the first full-length book on digital Bibles, theologian Jeffery Siker notes that YouVersion, in particular, "is clearly evangelical in scope, and proudly so," and he goes on to argue that the app itself promotes evangelical ways of thinking about scripture, its purpose, and the ways one should read it.[11] Digital religion scholar Heidi Campbell has argued that the new generation of tech-savvy Christians, whom she calls Digital Religious Creatives, are often surprised by the level of religious authority they generate through the things they make.[12] Similarly, Tim Hutchings has argued that evangelical developers who create Bible software have effectively, although unintentionally, employed social scientist B. J. Fogg's concept of "persuasive computing" to privilege evangelical readings of the Bible in their applications. YouVersion, he argues, prioritizes daily reading and study patterns that lead to evangelical conclusions about the text.[13] And yet YouVersion and other Bible apps did not come to exist in a vacuum. Instead, the apps and their creators are themselves shaped by a variety of factors, including evangelicalism's value systems around texts and technology as well as external factors like changes in screen technology, evolving social norms around the ever-present glowing rectangles in our pockets, and the technological values of individual church congregations.

This leads to the primary question of this book: What characteristics of evangelicalism have enabled it to create the most commercially successful and widely used Bible software? That is, what significance do the evangelical identities of developers and users have in shaping the feedback loop of Bible software creation and usage? What we will find, through interviews with the developers themselves and focus groups in evangelical churches, is that while Ochel and Brown may have viewed their plans to combine faith and ministry with business and media as novel, in fact, this impulse has deep roots in the history of the evangelical movement. As historian Timothy Gloege has shown in his analysis of the origins of Moody Bible Institute in the late nineteenth century, evangelicalism can be understood not only as a set of doctrinal beliefs or spiritual attitudes, but as a flexible outlook capable of adapting to economic and technological shifts and which sees a parallel between successful business outcomes and spiritual development.[14] As Bruce Shelley wrote in the 1960s, "Evangelical Christianity is not a religious organization. It is not primarily a theological system. It is more a mood, a perspective, an experience."[15] At times, this mood takes the form of conservative outlook that fears technology and its potentially negative moral influence, but other strands of evangelicalism readily employ technology in service of their greater mission, exemplified by Billy Graham, who embraced and mastered radio and television.[16] More recently, YouVersion has been so successful in understanding the power of modern technology that it was featured as a case study in behavioral engineer Nir Eyal's book *Hooked: How to Build Habit-Forming Products*.[17] In subsequent chapters, I will unpack this farrago of traits and argue that this flexibility and pragmatism regarding doctrine, practices, media, business, and politics are the very things that made evangelicals uniquely suited to be pioneers of the digital Bible industry and the primary early users of Bible software. I will show that the way evangelicals have embraced—and even dominated—both the production and consumption side of digital Bibles offers unique insight into the nature and meaning of the evangelical movement today.

Hopeful Entrepreneurial Pragmatism

The evangelicals studied in this book, both developers and end users, demonstrate the traditionally recognized emphasis on the importance of the Bible in their worship and personal spirituality. They also extend this with an expectation that regular Bible engagement will lead to spiritual change and pair it with a flexibility and openness about how they accomplish this goal. For the developers, this can be observed in the way they move fluidly between discussing business and technological success alongside spiritual and missionary success.[18] They do not equate the two, but they are comfortable with their being interrelated and intermingled on a parallel trajectory. In addition, while the companies under consideration have different business models and technological emphases, their outlooks on media, ministry, and business tend to share a set of common characteristics.

First, they have a *hopeful* outlook, exhibiting a net positive view of technology's potential for Christian ministry and personal growth. Although they are aware of the potentially negative aspects of modern technology such as distracting notifications and skim reading, they tend to believe it is better to be a part of technology development than to retreat from such a significant aspect of modern life. Second, they are highly engaged *entrepreneurs* and savvy business leaders capable of building successful technological and creative systems. Many of the successful Bible software companies were started by people with experience in the technology sector, and the evangelical penchant for integrating not only technology but also cultural trends and business methods served their companies well. Third, they are *pragmatic* in approach, making decisions based more on what "works"—in both moral and business senses—than on any systematic beliefs or direction from an authority. The company leaders seamlessly move between markers of spiritual and financial success, and they are willing to try almost anything as long as they can find data demonstrating its effectiveness. They take seriously the call to make disciples of all nations, and they are excited about the potential of technology for helping the church accomplish its mission. Together, these three traits can be combined into an attitude that we will call Hopeful Entrepreneurial Pragmatism (HEP), a summary of the approach evangelical software companies take toward the digital Bible.

It is a useful framework for understanding why evangelical Bible app developers have been so successful and why evangelical readers are so open to embracing these apps.

The Habits and Hermeneutics of Digital Bible Readers

Turning toward the end users of digital Bibles, we will see that evangelical readers bring with them the importance of regularly engaging in devotional and Bible study activities and the belief that they are to read the Bible for its capacity to "transform lives."[19] The Bible is not just a religious text for evangelicals, but a deep source of spiritual life and connection with God. The digital Bible gives them a variety of new means of accomplishing these goals and deepening this connection. From personalized reading plans to Bluetooth-enabled audio versions, the pragmatic flexibility (HEP) deeply engrained within evangelicalism means that they are open to all the options presented to them if it helps them become more like Christ. They use many different forms of media to access the Bible, including an array of hardware (smartphones, tablets, and computers), software (study apps, memorization apps, audiobooks), and printed books (journaling Bibles, study Bibles, reader's Bibles, etc.).[20]

They recognize that some of these are better for accomplishing particular types of Bible engagement (e.g., print for devotional reading, digital media for searches), and yet the churchgoers I studied admitted that their heuristic for choosing a Bible was often very simple. They chose what I like to call the NAB, or the Nearest Available Bible, which, due to the high percentage of smartphone ownership, is often a phone-based Bible app. This adaptable approach to both technology and the Bible then reinforces evangelical ideas about the nature of scripture and the goal of faith. And yet, as we will see, potential problems with screen-based media, such as increased distraction and decreased comprehension rates, at times undercut the perceived gains offered by digital Bible technology.

While basic factual comprehension is important in Bible reading, the spiritual experience one has while encountering the text may be even more central to one's faith. The data suggest that Bible readers tend to see a kinder, gentler God when they read about him on a screen and yet they report feeling more discouraged and confused by the encounter. Conversely, print readers tend to emphasize more of God's holiness and judgment, but report feeling more fulfilled and encouraged by the encounter. It is also important to recognize that these changes in the habits and hermeneutics of today's readers are not entirely comparable to the previous media shift to print. Where the print Bible entirely replaced the handwritten codex, today's Bible readers often use screen-based Bibles alongside a print Bible, and they add audio Bibles to the mix, creating a multimedia experience. This suggests a rich new environment for exploring the relationship between culture, scripture, and technology.

The State of the Digital Bible

In the summer of 2018, YouVersion celebrated the tenth anniversary of the release of its Bible app.[21] In the decade since its first release, YouVersion went from a small web-based experiment in user-generated content to being a staple in Apple's list of the top 50 free iOS apps, making it the most popular Bible app and a fixed part of the religious technological landscape.[22] But this landscape also extends beyond smartphone apps to include desktop Bible study software, Bible websites and tools, and social media where users share scripture with one another. In fact, the Barna Group's *State of the Bible* report found that from 2011 through 2021 approximately half of all Americans were "Bible users," meaning they engage the Bible in some form at least three to four times a year.[23] Of those Bible users, in 2019 more than half (55%) used the internet to access the scriptures, while others use smartphones (56%), apps (44%), audio versions (36%), and podcasts (36%).[24] These percentages continued to grow in 2020 and 2021 during the pandemic, before slowing in 2022. Overall, the trends indicate that in the forty years since 1982 when

Ochel and Brown released the first commercial Bible software, digital Bible usage has gone from 0% of all Americans to over 35%. For the remainder of the chapter, I will survey much of the work that has been done to document and understand these trends, highlighting gaps in the research that this book will attempt to fill.

In chapter 4, I will argue that Bible software development has taken place in four waves. These will be explored in more depth in that chapter, but here they will serve as a helpful outline for exploring the available research. In the first wave, (1) the preconsumer academic era (1950s–1970s), the only people using computers to interact with the Bible were scholars doing linguistic analysis. This shifted in the second wave, (2) the desktop era (1980s), when the first consumer applications were released. These applications, such as ThePerfectWord, Logos Bible Software, Accordance, and PC Study Bible, were designed primarily for study and exegesis and used by pastors, seminarians, and scholars.[25] Some authors released books such as *Bits, Bytes, and Biblical Studies* that cataloged how to use a computer for Bible study, and journals published reviews of new software releases, but there were not yet any large studies of users.[26] In the (3) internet era (1995 onward), websites like Bible Gateway and new translations like the NET Bible began appearing online, and their presence expanded digital Bible usage beyond the offices of pastors and linguists into the homes of regular Bible readers. This is also the era in which the study of digital religion began to emerge as a distinct discipline where scholars began to consider questions about media and religion broadly and the digital Bible specifically. Finally, in the (4) mobile era (2007 onward), marked by the release of the iPhone and YouVersion's Bible app, discussions about the digital Bible entered the mainstream as bloggers debated whether preaching with an iPad "sends an entirely different message to the congregation"[27] than a printed Bible or simply represents a natural transition in the digital age.

Bible App Usage Research

Beyond what has already been introduced above, the research into the growth and usage of digital Bibles can be roughly divided into four categories. First, some Bible software companies post summaries of their own analysis of their user's behavior patterns. Second, polling groups like Barna and the American Bible Society have conducted large-scale, longitudinal studies that examine how Americans interact with the Bible in both print and digital media. Third, several scholars have written essays that consider the influence of the digital Bible on overall religious readings, some of which were speculative, while others categorized and theorized more clearly. Finally, some sociologists of religion have examined small sample sizes, gathering qualitative and quantitative accounts of digital and print Bible usage as well as interrogating Bible software and the patterns of reading embedded therein. As we survey this growing body of research on the digital Bible, it will become clear that there is very little research on the religious identity of Bible software developers and the ways in which their evangelical leanings influence the development and features of the software.

The largest repository of information on the habits of digital Bible users is probably the information stored on YouVersion's servers that tracks how their users interact with the Bible app. YouVersion publishes some of this information in the form of annual infographics that give some indication of the amount of interaction the app receives from users.[28] It also releases snapshots on specific holidays that indicate the popularity of verses on days like Easter and Mother's Day.[29] Bible Gateway has also published information about patterns of usage related to cultural events. For example, after the deadliest mass shooting in modern US history in 2017, Bible Gateway and *Christianity Today* reported a surge in searches for comforting passages such as John 16:33 ("I have told you these things, so that in me you may have peace. In this world you will have trouble. But take heart! I have overcome the world") and Psalm 34:18 ("The Lord is close to the brokenhearted and saves those who are crushed in spirit"). They also found similar search spikes during eighteen other violent attacks.[30] Similarly, during Covid-19 lockdowns, YouVersion "saw searches increase by 80 percent in 2020," and the most-read verse was Isaiah 41:10, which begins, "So do not fear, for I am with you" (though, perhaps Isaiah 8:12 would have been more helpful).[31] Another window into online Bible usage comes from Stephen Smith, a developer at Bible Gateway, who posts analyses of data on his personal blog, *openbible.info*, including which verses are most popular on social media platforms and what Twitter users are giving up for Lent.

These snapshots offer interesting windows into the social patterns enabled by digital Bibles, but they are not able to consider how users move between their print and digital Bibles, or the role developers have in shaping and directing the media habits. The studies also tend to be limited to a particular platform or product.

Bible Reading Trends

The American Bible Society's annual State of the Bible research includes several trends that indicate that the number of Christians who use digital media to access the Bible has been steadily increasing over time, and they are doing so using many different forms of media.[32] In 2021, 59% of respondents said they preferred print overall, but in the eight years between 2011 and 2019, the number of American Bible readers who used a smartphone to access the Bible grew from less than a fifth (18%) to more than half (56%). This phone usage includes downloading and using apps on phones and tablets, which rose from 35% in 2015 to 44% in 2019, as well as other smartphone usage, including 60% of the users who reported that they "search on the phone" using a mobile web browser instead of a dedicated application. Nearly the same percentage reported "using the Internet" to access the Bible on a variety of other devices including laptops and desktops computers. Approximately one-third of Barna's respondents also reported other digital media usage, such as "listening to teaching about the Bible via podcast" (36%) and "listening to an audio version of the Bible" (36%).[33] Interestingly, ABS's data also indicate that women are more likely to have installed a Bible app and more likely to use it regularly, while men are more likely to listen to an audio version of the Bible.

In Barna's survey, after asking about different forms of media, they ask about the user's overall preference: "All things considered, in what format do you prefer to use the Bible—print, digital, or audio?" The data show that during the period of increased digital media usages, the percentage of participants who reported a preference for reading the Bible in print has consistently hovered around 75%. As one might expect, this varies by generation, with Gen Z being the first age group to rate print at under 50%.[34] The fact that three-quarters of Bible readers use digital media at least some of the time, but two-thirds still prefer print, indicates that there is room to explore the interplay between media and the settings in which one is preferred over the other, which I will do in chapters 6 and 7. These data also indicate that while print is still very important, the category of "digital Bible" is complex and multifaceted, and that the simple comparison of "print versus digital" does not tell the entire story of Bible engagement. Instead, print and digital should be understood less as a strict dichotomy than as a broad spectrum of Bible engagement experiences. This includes print and digital products that attempt to minimize distractions, such as printed Bibles without verse numbers, like the

ESV Reader's Edition, and apps that attempt to minimize distractions, like NeuBible. At the other end of the spectrum are products designed for a more research-oriented approach to the Bible, such as printed study Bibles and apps with commentaries and original language features. Later, we will hear from congregants who contrast the distractions of their phone with the level of concentration they can achieve with a printed Bible, but we will also hear from people for whom a printed Bible is much too inaccessible and complicated, while a digital Bible on a phone is more intimate and convenient.

Barna also examines general sentiments about the Bible, including why people say they read the Bible and their desire to read it more often. A striking 78% of Americans express a desire to read the Bible more often, including 20% of those who do not identify as Christians. As for those who recently began reading the Bible more often, more than half (56%) of them say it is part of their "faith journey," while others mentioned difficult life circumstances, significant life change, seeing the Bible change another person's life, or being invited to read it or attend a church with another person. Some evangelical commentators have suggested that this means readers are primary "therapeutically motivated,"[35] looking not for "truth" but for ideas that will help them enjoy their lives. And yet this desire for change connects with what we will see is one of the primary motivators of Bible software developers, particularly those at YouVersion. In the early 2010s, some respondents in Barna's surveys attributed their recent increase in Bible reading frequency to downloading a Bible app, suggesting that the novelty of Bible apps may have played a significant role in their behavior. This is an indicator that evangelical Bible software developers have, just by creating apps, influenced the way Americans engage with the Bible.

This book will attempt to fill a gap by unpacking the ways in which the developer's evangelical identities and views of scripture have shaped the process of creating Bible software and how that has, in turn, shaped the reading habits and hermeneutics of everyday readers who use it.

2
Studying Technology and Faith

In the last few decades, several academic disciplines have emerged with that each attempt to explain the fascinating and complex relationships between media, technology, and religion. This chapter will begin by tracing some of the early attempts to understand technology by media theorists and how those ideas evolved into more complex approaches such as social shaping of technology. We will then follow the emergence of the field of digital religion, a subfield of sociology of religion that draws together insights from religious studies and media studies. In what follows we will explore the insights these fields bring to understanding the development and evolution of Bible software and how they will contribute to studying the evangelicals who make these apps. If the reader is more interested in beginning to explore evangelicals and software development, this chapter could be skipped.

Early Approaches to Technology Culture

Before the development of digital religion as a discipline in the 1990s, scholars outside of religious studies began investigating the role of media and technology in society and developing theoretical approaches to address questions that arose from those inquiries. In the 1960s, Marshall McLuhan developed a set of theories around how media, technology, and communication influence human society and culture.[1] One of the key concepts in his work that continues to ripple through all subsequent media theory is that media themselves should be a subject of study, not merely the content carried by media. McLuhan went on to popularize the idea that the significance of the printing press was not that it allowed ideas (scripture, novels, pornography, etc.) to be spread more quickly, but that the printing press was itself a powerful idea, one that created a something he called "print culture."[2] In the mid-1960s, McLuhan's ideas were so popular that he made multiple television appearances and was covered in everything from the *New Yorker* to *Newsweek* to *Playboy*.[3] Two of his most successful students, Walter Ong and

People of the Screen. John Dyer, Oxford University Press. © Oxford University Press 2023.
DOI: 10.1093/oso/9780197636350.003.0002

Neil Postman, further developed McLuhan's ideas, turning them into a discipline called "media ecology," and their ideas were applied to understanding the significance of various Bible media over time.[4] Others, such as sociologist and philosopher Jacques Ellul who wrote about the importance, and indeed overwhelming, role of technology in shaping modern society, are sometimes folded into or appropriated by media ecologists.[5] However, in the 1970s, when many of McLuhan's predictions did not appear to be on the horizon, his star began to fade, and, simultaneously, thinkers from other disciplines began to see the need for alternative frameworks with more nuance and methodological heft.

The emerging discipline of philosophy of technology tended to see media ecology as a form of technological determinism that put too much weight on media as the sole driver of societal change, in much the same way that Marx's ideas have been considered economic determinism.[6] Writing in the early 1970s, John Fekete criticized McLuhan for not addressing the role humans play in technological development and use, arguing that by "denying that human action is itself responsible for the changes that our socio-cultural world is undergoing and will undergo, McLuhan necessarily denies that a critical attitude is morally significant or practically important."[7] Although statements such as "we shape our tools and thereafter they shape us"[8] indicate that McLuhan's views were somewhat more nuanced than his critics gave him credit for, some later media ecologists would embrace the term "determinism" but qualify it as a "soft determinism" that sees a greater role for transactional human interaction with technology.[9] If determinism was at one extreme in approaching technology and society, instrumentalism was considered to be its opposite, emphasizing that human agency is absolute and that technology is "value free," as philosopher Andrew Feenberg would put it.[10] Feenberg and others argue for the need to develop critical theories that take into account both the value-ladenness of media and the agency of humans.

Social Shaping of Technology

In the 1980s, a new approach called the social shaping of technology (SST) emerged out of science and technology studies as another counterresponse to technological determinism.[11] Technological determinism was seen as arguing that certain paths of technological innovation were inevitable, and, therefore, their effects on society were outside of human control.[12] SST advocates hoped to offer a fresh and more balanced means of interrogating the new wave of computer, medical, and weapons technologies that were being developed. SST approaches emphasize human and social agency and recognize that different cultures may adopt and adapt to technology in distinct ways depending on their existing value and authority structures.[13] They are more concerned with the organizational, political, economic, and cultural factors around the process of innovation than with the social changes brought by technology. Communication professor Janet Fulk writes that "a constantly evolving set of social structures and technological manifestations arises as groups selectively appropriate features of both technology and the broader social structure in which the group is embedded."[14] In the classic example of the bicycle, SST scholar Wiebe Bijker argues that what emerged from its social construction was not the "best" bicycle in an objective sense (i.e., design or efficiency), as if it were Plato's form of "bikeness," but the one that met the needs of the relevant social groups who had a stake in using the bicycle.[15]

Similarly, gender and technology scholars Cynthia Cockburn and Susan Omrud explore the development of the microwave in terms of gender values, arguing that it was initially conceived of as a masculine technology, but through social use, it was reconstructed as a feminine kitchen appliance.[16] Today, scholars continue to study gendered understandings of technology adoption, such as the way men and women differ in how they hold their phones.[17] In these cases, scholars from SST and other disciplines argue that technology does not develop "according to an inner technical logic but is instead a social product, patterned by the conditions of its creation and use."[18] Whereas deterministic approaches would look for inevitabilities within the technology itself ("the medium is the message"), SST emphasizes the choices available at each step in the process of innovation and see development as a nonlinear garden of forking paths.

Social Construction of Technology

Within the larger SST banner reside a number of models, and here we will focus on one called the social construction of technology (SCOT). The SCOT approach offers a clear methodology for approaching the development of a technology, and it is the tool I have chosen to draw on it as we examine at how Bible software evolved. Here we will explore five components of SCOT that Bijker and his colleague Trevor Pinch laid out in their initial model along with later adjustments that came through scholarly engagement.[19]

Their first concept is *interpretative flexibility*, which states that various groups of people will interpret differently, or assign different meanings to, a technological artifact. These interpretations may change over time, as has happened with the automobile, which in the 1890s was understood as "green" alternative to the pollution caused by horses in urban environments.[20] A second, closely related concept is *relevant social groups*, which include both the creators and users of the product as well as any subgroups (e.g., management and programming) or related groups (e.g., reviewers and government regulators). In the case of digitizing a religious text, a SCOT approach would identify users of various religious traditions, each of which might value some modes of interacting with their sacred texts over others. A Muslim group reading the Quran, for example, may interpret a user interface very differently than a Roman Catholic group reading a Bible. And where both of those groups might see the digitization of their sacred text as a boon to their faith, an atheist group might interpret the same digital Bible app as a powerful polemical tool that could be used to strengthen their arguments. Within each of these groups, one might also find interpretive differences between those holding positions of authority and the layperson within the group. In addition to these different types of end users, the SCOT approach would also identify other actors, such as the software company owner whose concerns may go beyond the spiritual into questions of financial viability and commercial success.

A third component of SCOT related to the first two is the *design flexibility*, which stresses that the creators of a technology have many choices to make in the creation of their products. There are always multiple ways to construct a tool, and any design one creates is merely a single point in a branching tree of possibilities. In the creation of a Bible app, this process begins when programmers create a user interface that requires them to make choices that favor one function over another. In the case of the digital Bible, these decisions have significant spiritual import because they privilege one form of Bible engagement over another.[21] Yet a social shaping approach also recognizes that programmers do not perform their work in isolation; users respond with comments, criticism, and economic choices that form a feedback loop leading to changes in the software and an ongoing cycle of feedback and software iteration.

Software developers can take these unexpected uses into account and adjust their products accordingly. This leads to a fourth component of the SCOT framework, *closure and stabilization*, which explains how the design process eventually settles on a product. Conflicts often arise about how a technology should be built or how it should function, but once all the groups have resolved the conflict, design and interpretive flexibility collapse, and the product reaches a state of closure and achieves stabilization. Pinch and Bijker provided two example mechanisms for closure. First, in rhetorical closure, the creators declare that there is no further problem and no additional changes are necessary. In software terms, this might be phrased "The bug has been fixed" or "The new feature is finished." The second closure mechanism occurs through redefinition when a developer responds to a user's complaint by declaring that it is actually solving a different problem. In the mid-2000s, most mobile phones, including popular models from BlackBerry, had a physical keyboard and many feature buttons. In 2007, Apple released the iPhone, which had a larger screen and a single physical button on the front. This brought closure and stabilization by redefining the problem in such a way as to value simplicity over other potential design routes.[22]

In the realm of digital Bible apps, Logos Bible Software's mobile applications for iPhone and Android initially displayed text on the screen in the form of pages that the user could swipe right and left, which mirrored the navigational metaphor in e-readers like the Amazon Kindle. Some users offered feedback indicating that they would prefer to have a scrolling interface where they could move up and down fluidly as they could in other products, such as YouVersion. Eventually, Logos made this an option in the software and announced it on the company's blog (rhetorical closure), which brought a form of stabilization to the platform by allowing it to support the preferences of multiple user groups.[23] YouVersion, on the other hand, still only displays one chapter at a time, which does not allow users to see the end of one chapter and the beginning of the next at the same time. Rather than changing this, YouVersion effectively engaged in a redefinition closure by emphasizing other features like reading plans and audio that are connected to a single chapter.

A fifth and final component of the SCOT approach is the *wider context*, which includes the larger sociocultural and political milieu in which the development process takes place. Political scientist Hans Klein notes that in Pinch and Bijker's original conception of SCOT, this played a smaller role in the process, but has since been expanded through critical work.[24] For digital

Bibles, the wider context expands our vision beyond the dynamic between users and developers, into considering other factors such as the hardware created by other companies on which the software runs. As we will see in future chapters, digital Bible software followed the trajectory of the technology industry from desktop computers to internet websites to mobile devices. With each of these technological shifts came new ways of interacting with the text as well as expectations and ideas about what might supplement the text. For example, in its initial iteration, the primary innovation of YouVersion website was its user-generated commentary, which reflected the growth of social media and crowdsourcing websites like Wikipedia at the time. This indicates that trends in the broader technological world play a role in the development of the digital sacred.

On the hardware side of digital Bibles, design professors Juhyun Eune and Minjeong Kang compared desktop, tablet, and phone screen sizes and concluded that different form factors were better suited to certain forms of Bible engagement than others.[25] Phones, with their smaller screens, are better suited for shorter periods of reading, and their constant availability makes them ideally suited for reminders, while desktop computers have more horsepower and screen real estate for studying. In this sense, the wider context can be seen as setting the parameters of design flexibility and influencing closure and stabilization.

Critiques and Expansion of SCOT

While the SCOT approach is helpful as a move away from technological determinism, it is not without its critics. Social scientist Langdon Winner proposed that SCOT was too narrowly focused on how technology arises without considering its consequences.[26] This concern is relevant to the present study because we are interested not only in the role of evangelical ideas in the formation of digital Bible technology, but also in the consequences of embedded evangelical values in a Bible app. Winner also argued that SCOT only focuses on the groups who have a role in shaping technology, not those without a voice who are nevertheless affected by technology. Here again, the millions of Bible app users who do not have a means of offering actionable feedback are nevertheless important subjects in the social shaping of Bible technology. In a related vein, Scheifinger, looking at the practices of Hindus online, has argued for the significance of what he calls "non-participatory digital religion" and shown that examining only the groups most involved in created a technology will overlook the role of other social groups in shaping and being shaped by a technology.[27] Other scholars criticized SST in a more general sense for its "excessive emphasis on agency and neglect of structure."[28] By "structure," Klein is referring not to the tool itself (i.e., a Bible app), but to the social structure in place around a technology, ranging from how consumers access and buy technology to the value systems of a culture that shape how individuals and communities adopt and negotiate technology. In addition to the structures that frame the agency of the development team, another key concern not addressed directly by SCOT is how the technological product constructs a context that imposes a boundary around what is possible for the user. In the case of a Bible app, the features and interface structure the kinds of engagement that are possible. As consumers, users are free to choose from different Bible apps, but once inside the app, they are only free to do what the developers have created and prioritized.

These critiques led scholars in the field to expand on the original four components, including Bijker himself, who added another component to his original SCOT proposal that he called the *technological frame*, or the "frame with respect to technology."[29] This component attempts to address criticisms of early versions of SCOT that said that its approach to groups was too simplistic and did not account for existing beliefs, theories, prejudices, taboos, and other structural considerations that would tend to promote some technological innovations and discourage others. The technological frame is related to the concept of the wider context, but it focuses on what emerges from within a group and in the social structures between groups. This includes the ways both technology creators and users are positioned toward the social, political, and economic forces around them. In the realm of religious belief, it also includes the authority structures, tradition, and beliefs of a particular group that shape how it negotiates and adopts new technology. As we will see below, some Christian traditions have resisted adopting certain technologies such as digital Bibles in a worship service, while others, such as evangelicals, demonstrate a technological frame that is open to new media in their personal and corporate religious practices. While SST approaches like SCOT offer helpful methodological guidance for understanding digital Bibles, the concept of the technological frame points to the need for a more specific and nuanced approach to the development and adoption of technology within religious contexts. We find this approach in the area of digital religion and in scholars like Heidi Campbell, who have proposed a religious social shaping of technology. In addition, the SCOT approach is useful for analyzing the formation of a technology, but the traditional tools of sociology for fieldwork will be necessary to investigate the impact of the technology once it has been formed.

Digital Religion

The above sketch has explored the ways that scholars from various nonreligious disciplines have examined the relationship between media and culture, and we now turn to the more specific study of media and religion. In the 1950s, publishers began using computers to work with biblical texts to generate Bible indexes (see chapter 4), but these early efforts garnered little attention from religious studies scholars. However, in the 1980s, as the personal computer spread and early network pioneers dialed into bulletin boards and database services, scholars quickly moved beyond technical discussions and began to consider other aspects of computer users' identity, including their religious practices and debates.[30] By 1990s, scholars began to use terms like "cyber-religion" to describe the practices of faith that began to emerge on these network-connected computers.

Some media scholars like Lorne Dawson used the term "cyber-religion" exclusively to describe "those religious organizations or groups which only exist in cyberspace,"[31] while others like Brenda Brasher defined it more broadly as "the presence of religious organization and religious activities in cyberspace."[32] In the mid-1990s, the final restrictions on commercial internet traffic were lifted, and more traditional religions began to find their way online, and by the mid-2000s, the concepts around "cyber-religion" reached an endpoint, and the subfield of "digital religion" emerged as a distinct academic discipline.

Since its origin in the 1990s, digital religion has developed and matured from a largely descriptive and speculative exercise to become a more theoretically robust and grounded approach. Sociologists Morten Højsgaard and Margit Warburg used the metaphor of waves to describe a consecutive and overlapping series of approaches to studying religion and the internet. They argued that the first wave tended to focus on (and speculate about) the new and extraordinary possibilities of the online world, the second wave became more realistic in its framing and categorization of what was happening, and the third wave introduced new theoretical and interpretive frameworks.[33]

A decade later, Campbell and her colleague Brian Altenhofen would label these first three waves "the descriptive," "the categorical," and the "theoretical."[34] They added fourth wave, the "integrated/convergent," to which they sought to contribute. This research in this book is positioned within this multidisciplinary, integrated/convergent way of understanding technology and religion, and because each wave draws on and learns from the previous waves, it is helpful to recount their development and maturation.

An example of the first, "descriptive" phase is computer columnist Mark Kellner's book *God on the Internet*. At a time when Google was still a research project, Kellner cataloged and indexed all the religious websites he could find.[35] At this stage, religion on the web was in a nascent and evolving stage, and because of this, he and other authors tended to focus on reporting what was happening and offering guidance on how to join the movement, but their work did not provide much in the way of reflective analysis. Religious scholars were not far behind, offering more comprehensive descriptions and surfacing questions about what might happen as the internet adoption increased. Communications professor Stephen O'Leary wrote one of the first such accounts, drawing on the work of media ecology and communication theory including the work of Ong, to "speculate on the transformation of religious beliefs and practices as these are mediated by new technologies."[36] O'Leary continued his Ongian analysis of online religion with Brenda Brasher, analyzing the ways in which online "speech" functioned and the place of online religious texts, including the Bible, in those acts.[37] Other scholars began to ask questions about how the Bible might change as it moved from paper to pixels and how those changes might affect authority, canonicity, and the understanding of the Bible as an image or "artifact with symbolic freight."[38]

Theologian Tom Beaudoin, writing as the internet was reaching mainstream adoption, wondered if the free interconnected flow of information on the internet would make it impossible to maintain a fixed biblical text or an authoritative interpretation. He argued that "reading the scriptures has always been hyper-textual; cyberspace just helps us see that more clearly and might give us new options for interpretation."[39] Some bloggers have, in fact, called for the ability to "build your own Bible," and study-oriented applications encourage users to click through to additional resources as Beaudoin suggested.[40] However, all of the major Bible applications still include a fixed, canonical version of the biblical text, and none include the ability to reorder the canon or edit the text of the biblical books. Social shaping approaches to technology would suggest that even when a technology can do something (edit the canon), this pathway is not inevitably determined if the relevant social groups (in SCOT terms) do not desire the feature. In the case of digital Bibles, the primary or dominant group has been evangelicalism, and therefore, one might expect that its design flexibility would be governed by its conservative understandings of scripture and canon. If a user from another religious tradition suggested a heavily redacted Bible such as the one created by Thomas Jefferson, an evangelical Bible developer might respond with rhetorical closure saying, "The canon is a feature, not a bug."[41]

Moving into the "categorical" phase, sociologists of religion Jeffrey Haddon and Douglas Cowan's edited volume was one of the first works to introduce systematic thinking about religion and the internet, with authors offering an array of new methodological approaches and example analysis.[42] Christopher Helland's essay arguing for a distinction between "online religion" and "religion online" was particularly influential.[43] He used the term "religion online" to describe the ways in which traditional religious hierarchical structures were utilizing the Internet as a tool to further their aims. He argued that these organizations tended to present noninteractive information in a top-down vertical manner that preserved existing authority, control, and status. In contrast to this, Helland used the term "online religion" to describe what he saw as a new form of religious praxis that reflected "the configuration of the Internet medium itself."[44] He saw the internet as inherently nonhierarchical and argued that its nature as open-ended allowed these new forms to emerge in a "bottom up" manner. A related typology came from Cowan, who suggested applying the terms "open source" and "closed source," from the programming world, to religious practices. Just as closed-source programming is controlled by the developer, and end users have no input or rights to change it, closed-source religion describes hierarchical religious organizations that do not tend to incorporate parishioners' ideas into official dogma. In contrast, anyone can use or contribute to open-source software, and Cowan suggested that open-source religions are those that encourage innovation from noninstitutional sources.

Scholars, including Helland and Cowan themselves, would later point out that these distinctions are somewhat murky and cannot be strictly maintained.[45] For example, it is difficult to know if reading a religious text online (i.e., a digital Bible) can be cleanly categorized as "online religion" or "religion online."[46] And yet, despite their limitations, Helland's classifications allow us to distinguish between digital Bible *creation*—by the Christian organizations that generate Bible translations and license it to software developers who in turn create applications that present the authoritatively determined text to users—and the use of digital Bibles for various *practices* such prayer, devotional reading, and small-group interaction. There are also some open-source Bible applications that appear to be a rather literal application of Cowan's open-source religion, except that they, too, generally use fixed translations of the Bible and do not include features that would allow users to modify the text. An app that subverts traditional religious authority structures is Our Bible, which was released in 2018 and whose "goal is to untangle the binds that Christian colonizers have spread across the globe over hundreds of years"; it does so through "devotionals highlighting pro LGBT, pro-women and [encouraging] interfaith inclusivity."[47] But the fact that the app only contains recognized translations suggests that traditional religious authority structures are still very much in operation, circumscribing what constitutes "the Bible" in print and in digital forms. In these examples, "religion online" still appears to override "online religion."

Højsgaard and Warburg suggested that, as they were writing in the mid-2000s, a third wave of research was emerging that was moving from broad categorizations toward more nuanced theoretical and interpretive approaches that incorporated theories from other fields into a diverse "bricolage." Scholars began to refine the techniques of ethnography and develop new approaches, such as journalism professor Robert Kozinets's "Netnography" for studying social media.[48] Thy began to recognize that the internet was becoming a part of everyday life and the distinctions between online and offline practices beginning to dissolve as users moved fluidly between the two. This led them to explore themes like community, authority, and ritual. Some questioned the efficacy of online "pseudocommunities,"[49] while others offered a rich exploration of how online churches understood the relationship between digital religion. and everyday life, and Campbell further explored the ways in which the internet as a social network also enabled it to be a spiritual network for some religious people.[50] Scholars also explored how the internet could allow new religious authorities to emerge and challenge existing structures, and could simultaneously enable existing religious leaders to exert their influence, in cases such as ultra-Orthodox Jewish communities online, Muslim podcasts, Buddhist web discussion boards, and Roman Catholic forums popular in Poland.[51]

These questions of authority are of particular interest related to evangelicals, who often reject religious structures and traditions in favor of making the Bible itself their ultimate authority, all the while using Bible and religious study aids, both offline and online, created by religious leaders within the evangelical umbrella. Many American evangelicals worship in nondenominational churches or interact with parachurch organizations, neither of which have the kind of outwardly visible hierarchical, ecclesiological structure found in Roman Catholicism, Protestant denominations, or other religions like Islam or Judaism. And yet, while the authority structures are formally decentralized in evangelicalism, we will see that the development of digital Bible software has taken place within a tightly woven network of evangelical institutions, ministries, churches, and businesses. Some of these churches and ministries grow so large that they function with the power and influence of a denomination in a previous era.[52] This may explain why digital Bibles are not sources of canon remixing, as some early scholars suggested they would be.

Religious Social Shaping of Technology

This leads us to Campbell's proposed fourth wave of digital Bible research, as well as her religious social shaping of technology model that applies the SST models discussed above to the study of digital religion. This fourth wave began in the 2010s as smartphones became more common and the internet became more embedded in everyday life, including social interactions on social media, economic transactions through online shopping, and work-related tasks through telecommuting. Scholars argued that a hard distinction between online and offline could no longer be maintained, and they began examining the relationship between the two and how the boundaries and interactions in one mapped to and informed the other. In this fourth wave, Campbell and Lövheim presented new typologies for categorization and interpretation, stressing the need to "push for reflection on the social and institutional aspects of practicing religion online,"[53] which seems to parallel the later additions to social shaping and construction models. These methods continue to be developed, including Hutchings's work, which specifies several approaches to studying religious apps such as surveys and interviews, some of which will be employed in this study.[54]

Campbell's religious social shaping of technology (RSST) offers an additional layer of analysis for technology and religion. Although she and others continue to refine the model to fit within other areas of study, the core approach includes recognizing that religious groups rarely fully embrace or fully reject new media, but instead undergo a series of dialectical processes in which a group's beliefs, identity, and structure influence how they negotiate the adoption of the technology.[55] Campbell's RSST proposal includes four layers of investigation: (1) history and tradition, the ideas about community, authority, and previous media that might inform how new technology is understood; (2) core beliefs, the key religious and social values or dogmas that undergird the practices of a community and shape its responses to new things; (3) negotiation processes, how the community responds to new technology, including what it accepts and what it modifies or rejects according to its traditions and beliefs; and (4) communal framing, how community leaders and members talk about new media through official and causal channels.[56]

Some of the examples Campbell gives have strong authority structures, such as Amish, ultra-Orthodox Jews, and Muslims, where the negotiation is ultimately realized by a few powerful individuals within the community. However, she also considers Christian use of the internet more generally, where decisions are not dictated by a small leadership team but transmitted more broadly. Evangelicals, Campbell argues, are a prime example of a religious tradition with an "accept and appropriate" outlook on technology, especially tools that can complement its values: "From the printing press onwards using media technology to facilitate mass evangelization has been a marker of evangelical spirituality and practice."[57] Campbell and her collaborators analyzed and categorized research methods for digital religion studies, distinguishing between digital environments, tools, and frames.[58] I will deploy the RSST approach more directly in later chapters, discussing how evangelical Bible readers understand the significance of the digital Bible and make choices about how, when, and where to adopt it or reject it.

Study Groups and Approach

In the coming chapters, we will put the social shaping of technology (SCOT) and religious social shaping of technology (RSST) approaches into action as we study the creation and evaluation of Bible software. In SCOT terms our two relevant groups are Bible software developers and Bible software users, and in RSST terms we need to examine their history and beliefs and how they have tended to approach technology. The following chapter will look at Bible software through the lens of Hopeful Entrepreneurial Pragmatism, arguing that it is the "technological frame" (SCOT) or "communal frame" (RSST) for evangelicals. After establishing the evangelical approach to technology itself, I identified several Bible software companies to study and several churches in which to conduct focus groups.

My initial list of companies included the following: Accordance, Bible Analyzer, Bible Gateway, BibleWorks, Crosswire, Digital Bible Society, e-Sword, ESV online, Logos Bible Software, Glo Bible, OliveTree, PC Study Bible, QuickSearch, SwordSearcher, WordSEARCH, and YouVersion.[59] In chapter 4, I will explore the origins of several of these companies and their deeper connections to evangelicalism and its institutions, but for interviews, these needed to be whittled down to a representative and useful sample. I selected YouVersion because it has been one of the most often used Bible apps since the inception of Apple's App Store and because it is entirely supported by donations from its hosting church (Life.Church, formerly LifeChurch.tv), partner ministries, and its own users.[60] Next, I selected Bible Gateway because it has a large audience of users and because it represents a different kind of Bible software with a business model different from YouVersion's. Bible Gateway is currently owned by Zondervan, a large US-based Christian publisher that is in turned owned by HarperCollins Publishing.[61] Zondervan is also the US publisher of the New International Version (NIV), the Bible version most widely used by evangelicals. For many years, BibleGateway.com was one of the few places where one could access the NIV online, which led Bible Gateway to become one of the most highly trafficked Christian websites. It continues to be one of the first websites to be displayed in search engines when one searches for biblical texts. This leads to Faithlife, the company that produces Logos Bible Software and calls itself "the worldwide leader in electronic tools and resources for Bible study."[62] Logos was selected because it represents a more academically oriented application, with features designed for studying the original languages, accessing additional secondary literature, and preparing teaching materials.[63] Its mobile version was released the year following YouVersion's first release, and while YouVersion focuses almost entirely on Bible reading and Bible Gateway offers some purchasable content and notes, Logos offers a full array of scholarly materials and resources, making it a profitable software company.

Together, these three companies hold a significant market share, represent a range of business models, and emphasize different kinds of Bible engagement, making them an ideal representative sampling of digital Bible software producers. In addition, as we will see in chapter 4, these three companies represent three waves of Bible software development, the desktop era of the 1980s in Logos Bible Software, the internet era of the mid-1990s in Bible Gateway, and the mobile era of this century in YouVersion. I contacted each of these companies and requested a one-hour interview with developers and business leaders, and these interviews form the basis of chapter 5.

To study the end users, I selected three churches in the Dallas / Fort Worth area representative of evangelicalism and conducted focus groups within them. I have given each of churches in this study a pseudonym: City Bible Church, a twenty-five-hundred-member "multiethnic" church located in what *Forbes* magazine called the "most diverse neighborhood" in the United States;[64] Petra Community Church, a nondenominational church with approximately five thousand in weekly attendance and whose pastor has been recognized as one of the most influential and well-known preachers;[65] and Hidden Baptist Church (HBC), which recently changed its name from "First Baptist Church" in an attempt to appeal to a wider range of evangelicals who are not inclined toward denominations, part of a trend among some churches in the Southern Baptist Convention.[66] HBC draws its doctrinal emphasis from the 1963 Baptist Faith and Message. While City Bible and Petra Community have their own doctrinal statement, each puts the Bible as the first and foundational doctrine of its faith statement.

Each church allowed me to enter several of the Sunday morning or evening group meetings where I gave out a Bible engagement survey and conducted a Bible comprehension examination. I considered several biblical texts for the comprehension examination, but I settled on the Epistle of Jude for the following reasons. First, of the canonical epistles, I could find no references to Jude in the sermons or study materials of the churches' website, indicating that it is not often read or studied and that the participants were relatively unlikely to know the answers ahead of time. Indeed, scholars have noted that Jude is "rarely the text for a sermon"[67] and that its message is "alien to many in today's world,"[68] making Jude perhaps "the most neglected book in the New Testament."[69] Second, Jude is very short (approximately 650 words), which would mean that there was little chance that the difference in print versus digital reading would be due to fatigue in scrolling. Third, while it takes the familiar form of epistolary literature, Jude references several biblical narratives, enabling me to ask readers about the order of those stories. Finally, Jude contains both strong warnings ("Jesus . . . destroyed those who did not believe" [Jude 1:5]) and more gentle admonitions ("Be merciful to those who doubt" [Jude 1:22]) that I hoped might trigger emotional responses that might differed across media.

I also asked my subjects to participate in a ten-day Bible reading plan in the Gospel of John. I chose John because it had a prebuilt reading plan in YouVersion, was longer than a week, but was not so long that users might quit. In both the comprehension test and daily reading plan, half the group used their phones and the other half read in their printed Bibles. Finally, I asked the groups questions about how they use and understand the Bible. In all, around six hundred people across six groups and three churches participated in the study. Their responses will be analyzed in chapters 6 and 7.

Together, the focus group interviews with three software development teams and focus group assessments with three churches let us explore both sides of the digital Bible industry. These tools will allow us to observe the ways evangelical churchgoers use and discuss digital Bibles and uncover how the evangelical identities of the developers have contributed to the social shaping of Bible technology. But the first step in both the SCOT and RSST approaches is to understand evangelicalism itself, not only what evangelicals say they believe, but how they tend to react to cultural and technological change.

3
Evangelicals, the Bible, and Technology

In 1942, the National Association of Evangelicals (NAE) was founded as a kind of reboot of American evangelicalism under the banner of neoevangelicalism. However, in 2017, the organization noted that the most common reason people visited its website (nae.net) was to find a definition of the term "evangelical."[1] The NAE highlighted this pattern to make the point that while the term "evangelical" is used often in discussions about religion, it is not often well understood or clearly circumscribed. Indeed, the challenge in defining the term reaches back into the eighteenth century when Lord Shaftesbury, writing in the British context, lamented, "I know what constituted an Evangelical in former times, [but] I have no clear notion what constitutes one now."[2] In early 2016, the definition came under a fresh wave of scrutiny when 81% of Americans who identified as "white evangelicals" voted for Donald Trump,[3] a candidate whose public image stood against what had once been considered essential to evangelical morality. This quandary over defining evangelicalism is important because we are exploring why American evangelicals have been so drawn to the digital Bible, how its development and consumption have influenced the way they engage the Bible, and how this can inform us regarding the current state of evangelicalism.

Indeed, the first step in employing a social construction of technology (SCOT) approach is mapping out the "relevant social groups" and their "technological frame." This involves considering the beliefs and values of those groups and exploring how they contribute to the way each group assigns meaning to technological artifacts.[4] In our case, the overarching group is American evangelical Christians with a focus on two subgroups: digital Bible developers and Bible readers in a church setting.[5] This means that a deeper investigation into these groups will require us to unpack the historical development of the term "evangelical," in part because of the aforementioned disputes over the boundaries of the label, but also because a closer look offers us additional insight into how evangelicalism as social grouping has tended to interact with technologies. Moving from the general SCOT approach into Campbell's more focused religious social shaping of technology approach,

People of the Screen. John Dyer, Oxford University Press. © Oxford University Press 2023.
DOI: 10.1093/oso/9780197636350.003.0003

one should work to identify the history and traditions, core beliefs and patterns, community negotiation, and community framing and discourse of the religious group.[6] The important methodological principle is that a group's response to technology is "negotiable," not predetermined, and that a group's unique negotiation process is born out of its values, including its history and religious tradition, as well as the ways in which it has negotiated technological change in the past.[7] Religious groups like evangelicals often gain definition and clarity in their interactions with the culture around them, and technology can serve as a window into this relationship.

In this chapter, I will explore several avenues for understanding the characteristics and historical development of American evangelicalism, especially its emphasis on the Bible, and I will then offer a description of its technological frame and negotiation approach that I have labeled Hopeful Entrepreneurial Pragmatism (HEP). I hope to show that although scholars have suggested several different and sometimes conflicting ways of understanding evangelicalism—and some such as historian Kenneth Stewart argue that it is better to speak of multiple overlapping "evangelicalisms" rather than a single coherent group—the story of the digital Bible brings together characteristics shared by several approaches.[8]

Exploring Evangelicalism

Methods of Approach

At the turn of the millennium, Timothy Weber argued that "defining evangelicalism has become one of the biggest problems in American religious historiography,"[9] and this problem continues today across many disciplines. Because of evangelicalism's complexity, we will take an integrated approach that draws on historical, doctrinal, phenomenological, and sociological accounts, which together offer a complex, nuanced, and rich understanding of evangelicalism. The historical approaches seek a clear starting point for evangelicalism as a unique religious movement, while sociological approaches tend to focus on the strands of evangelicalism that emerged in the later part of the twentieth century. The doctrinal definitions that evangelical leaders themselves have offered can also be contrasted with the polls that examine the behaviors of people who self-identify as "evangelical" or "born-again." Across this spectrum, one of the recurring questions is whether evangelicalism is best understood through internal theological language or external descriptions of characteristics and behaviors that, in some cases, do not appear connected to religious belief. Writing from the UK, Samuel Crossley puts the question this way: "Are Evangelicals primarily recognizable by the doctrinal propositions to which they actively subscribe, or by observable and phenomenological traits which they may or may not consciously determine?"[10] Rather than take a single approach, we will draw from each of these perspectives, noting their strengths and limitations, and forming a richer view of evangelicals and their approach to the digital Bible.

We begin with a broad outline of the history of evangelicalism, drawing on scholarship that highlights evangelicals' relationship to the Bible and its posture toward societal change. My goal is not to give an exhaustive account of evangelicalism's historical development, but rather, by looking at three eras—the early years of the Reformation, the turn of the eighteenth century, and the mid-twentieth century onward—to see how common elements of evangelicalism's history and tradition bubble up through time, eventually setting the stage for the Bible software that would emerge in the 1980s.

Early Protestant "Evangelicals" (1517)

In the early Reformation, the term "evangelical," derived from the Greek *evangelion* (εὐαγγέλιον), meaning "gospel," was first used as an alternative to the term "protestant." If the term "protestant" focused on what the Reformers were against (the Roman Catholic Church), Martin Luther used the Latin word *evangelium* to show what the emerging movement was for: the gospel, a simple but powerful message of salvation through faith in Jesus found in passages like 1 Corinthians 15:1–5 and Ephesians 2:8–9.

The term made its way into English in the early sixteenth century, when William Tyndale began using word "evangelical" as an adjective, and the Roman Catholic Thomas More picked up the term, referring negatively to "Tyndale [and] his evangelical brother Barns."[11] For More, "evangelical" was a pejorative synonym for Protestant. But historian Lindford Fisher has argued that in this era, the term began to take on a new sense, namely that to be an evangelical was to be a "true Christian" that was more "gospel-centered, Bible-based, and authentic"[12] than not only Roman Catholics, but also other Protestants. Over time, just as Protestants had distanced themselves from the problems they saw in Rome's understanding of the Bible, evangelicals began to emerge as those who distanced themselves from the problems they saw in how other Protestants interpreted the Bible.

For all the changes that evangelicalism would undergo in the coming centuries, this focus on a gospel message that is "Bible based" would become a key marker of evangelical identity. The central importance of the Bible in faith and practice would also become an important factor in the emergence of the Bible software industry as evangelical programmers sought to combine their devotion to the Bible with their technological skill.

The Emergence of Evangelical*ism* (1700s)

The challenge in more narrowly defining evangelicalism as something distinct from Protestantism emerges in the late seventeenth and early eighteen centuries. In historian David Bebbington's classic work, he argues that a new form of Christian practice emerged in the 1730s in England with the conversion of Howell Harris and Daniel Rowland, followed by George Whitefield and the Wesley brothers, who spread the movement in the American colonies alongside Jonathan Edwards. Bebbington identified four characteristics of the new movement: "*conversionism*, the belief that lives need to be changed; *activism*, the expression of the gospel in effort; *biblicism*, a particular regard for the Bible; and what may be called *crucicentrism*, a stress on the sacrifice of Christ on the cross."[13] These four characteristics were not necessarily exclusive to evangelicals, but Bebbington argued that their combination was unique. In particular, he points out that the "activism" was something new, especially as it took form in the missions movement of the nineteenth century. This impulse to spread the gospel using nineteenth-century methods and technology, such as faster ships, cheaper printing, and better hospitals, would continue in the twentieth century as evangelicals embraced radio and television, and later computers, the internet, and mobile phones to share the scriptures.[14] As we will see later, the hopeful embrace of technology also incorporated a pragmatic approach that valued missional outcomes, sometimes at the expense of doctrinal conviction.[15]

While Bebbington places evangelicalism's origins in the 1730s and his thesis has "achieved scholarly hegemony in less than ten years after its initial publication,"[16] others find the origins of the movement somewhat earlier.[17] Journalist and historian Molly Worthen writes, "In the late seventeenth century, Pietist preachers critiqued the state churches that emerged from the Reformation as overly formal and cerebral. They called on believers to study the Bible and strive for personal holiness."[18] Similarly, historian Mark Noll's work seeks to push "the history [of evangelicalism] back into the seventeenth century"[19] so as to include a more international group of Christians who displayed characteristics of evangelicalism.[20]

Although these scholars disagree on the exact start date, they agree that the movement emerged, not only as a distinct set of religious ideas, but as a response to massive societal changes. In her biographical treatment of Sarah Osborn, an influential American evangelical, religious historian Catherine Brekus writes that "Sarah was drawn to evangelicalism because it helped her make sense of changes in everyday life that did not yet have a name. Words like *capitalism*, *individualism*, *Enlightenment*, and *humanitarianism* were not coined until the late eighteenth and nineteenth centuries, but language often lags behind reality."[21] In other words, evangelicalism was as much a distinct set of religious ideas and practices as it was a posture for absorbing and simultaneously resisting what was happening societally. In a similar vein, historian Kristen Kobes Du Mez traces evangelicalism's bent toward "muscular Christianity" in part as a male response to the perceived loss of manhood that came in the move toward cities and desk work.[22] In such works, a portrait of evangelicalism begins to emerge that is simultaneously a set of religious beliefs and practices alongside a distinct disposition, initially toward the Enlightenment and later toward other developing social realities, sometimes perceived as threats and sometime as opportunities.

Neoevangelicalism (1940s)

If evangelicalism initially emerged as a distinct movement in the post-Enlightenment world, another significant era of evangelicalism—what theologian Alistair McGrath called "an evangelical renaissance in the West"[23]—began during the middle of the twentieth century. Prior to this, evangelicalism had become largely synonymous with fundamentalism, or those who held to a set of fundamental doctrines.[24] By the 1930s, some conservative evangelicals in North America felt that fundamentalism had lost its way, taking negative stances such as "anti-intellectualism that suspects scholarship and formal learning . . . apathy toward involvement in social concern . . . [and] separation from all association with churches that are not themselves doctrinally pure."[25] Some of these men, including Billy Graham, Harold John Ockenga, Carl F. H. Henry, and Charles Fuller, began to use the term "evangelical" not as a synonym for fundamentalism, but a way of distinguishing their outlook as a more authentic form of faith.

This reframing recalls Fisher's argument that evangelicals tend to see themselves as a form of "true Christianity." The neoevangelicals went on to embrace the media and technology of their day, creating radio programs, magazines, and book publishers, as well as universities and seminaries. At the same time, their self-identification as "evangelical" kicked off renewed interest in more clearly defining the term and its boundaries. Crossley argues that beginning in this era, the word was used with "greater intensity and was deployed in an increasingly technical manner."[26] But while evangelicals attempted to define themselves as adhering to a set of conservative doctrines and outward practices, historian of evangelicalism George Marsden argues that "American evangelicalism in the 1960s was a vast, largely disconnected conglomeration of widely diverse groups."[27] This decentralized but interconnected state of the movement betrays another key aspect of American evangelicalism—its deeply independent and entrepreneurial character. As Protestants, evangelicals tend to reject formal religious authority structures, and as Americans, they could not rely on the government funding found in Europe. This led American evangelicals to be highly adaptable and comfortable with creating new models of ministry based on cultural trends. In the twentieth century, as evangelicals embraced contemporary media such as television, radio, and print, some began to form small religious empires around personalities and programs. Though not formally connected through an ecclesial structure, these churches and ministries fed off and relied on one another, and, as we will see in the next chapter, these organizations formed the strata on which evangelicals with technical knowledge would build the digital Bible industry.

In the mid-1970s, evangelicalism began to take on a more political meaning when Jimmy Carter, a self-proclaimed "born-again" Christian, was elected as the president of the United States, and *Newsweek* declared 1976 to be the "year of the evangelical."[28] *Time* magazine identified evangelicals as "basically conventional Protestants who hold staunchly to the authority of the Bible . . . [and] in making a conscious personal commitment to Christ . . . known as the born-again experience."[29] This public attention and discussion led sociologists like Stephen Warner to call for a fresh look at evangelicalism, acknowledging biases and barriers within his discipline.[30]

He concluded that evangelicalism should be understood as a "*Biblically-based* Christianity that emphasizes the *personal relationship* of the believer to Jesus."[31] In the early 1980s, other sociologists took up the charge offering more careful analyses of conservative Protestantism and evangelicalism.[32] James Hunter, for example, expanded Warner's understanding of a personal relationship with Jesus to include affirmations of the deity of Christ and the necessity of personal faith.[33] He also expanded the idea of being biblically based to include the concept of inerrancy, but Nancy Ammerman countered that including inerrancy narrowed the definition too much and excluded the "growing segment within evangelicalism that would not claim that they believe the Bible to be 'inerrant.'"[34]

Historians like Bebbington would draw on this discussion as they worked to connect the neoevangelicalism of the twentieth century with earlier iterations of evangelicalism, all of which emphasized the importance of the Bible. Although Carter's presidency brought the term "evangelical" into the mainstream, some evangelicals such as Jerry Falwell Sr. saw more opportunity for political gain through Carter's political opponents, the Republican Party. In the late 1970s, Falwell and others formed the New Christian Right and the Moral Majority, which some in the media credited with helping Reagan win his first presidential election.[35] Although some sociologists disputed the claim that conservative Christians affected the 1980 election, the "new religious Right," as they were called at the time, continued to grow in power as they mapped their theology to the Republican Party's ideals.[36]

This increased presence in public life led political polling organizations to begin including questions about whether or not participants had been "born-again" or had a born-again experience, and researchers used this information to study evangelical behavior, especially during elections. However, Stetzer points out that these polling instruments varied during the 1970s through the late 1990s, and "these definitional variances . . . created a disparity in percentage of evangelicals studied, ranging from 7 percent (Barna, 1998) to 47 percent (Gallup, 1999)."[37] Stetzer goes on to argue that "evangelical" should not be understood by self-identification with the term but by adherence to core beliefs that roughly correspond to the Bebbington Quadrilateral. LifeWay Research and the Barna Group now use these theological indicators in some of their research. As the methodology stabilized by the start of the twenty-first century, Noll notes, "Evangelical Christianity had come to constitute the second largest grouping of Christian believers in the world."[38]

The Pew Research Center reports that 25.4% of Americans identify as evangelical, but that the percentage varies by metro area, falling as low as 9% in New York City and reaching as high as 38% in Dallas, Texas.[39] These statistics have been shifting in the last decade, with the Public Religion Research Institute (PRRI) reporting that during the decade spanning from 2006 to 2017, those identifying as white evangelical Protestants dropped from nearly one-quarter (23%) to fewer than one-fifth (17%).[40] PRRI also notes that other significant demographic changes have taken place during this same time period, and that while the number of white evangelicals has decreased, the number of nonwhite Americans who identify as evangelical has increased, even as the term "evangelical" has come to be associated with white culture.

While the demographics are shifting for evangelicalism as a whole, polls indicate that "white evangelical Protestants" vote consistently for Republican candidates.[41] This voting pattern has led some to contend that evangelicalism in America no longer has a theological or doctrinal character but has become largely synonymous with political positions. Mark Labberton, president of Fuller Theological Seminary, offered this lament:

> The word "evangelical" has morphed from being commonly used to describe a set of theological and spiritual commitments into a passionately defended, theo-political brand. Worse, that brand has become synonymous with social arrogance, ignorance and prejudice—all antithetical to the gospel of Jesus Christ.[42]

Labberton was writing in response to the election of Donald Trump, whose presidency led to a fresh wave of evangelicals battling over both the usefulness of the term and who should be included in it.[43] Analysis and interviews about why evangelicals voted for Trump suggest a wide range of reasons, including that they felt he would represent their religious ideals in government, such as appointing conservative judges, and that he would advocate for their religious freedom. At the same time, some explained his immoral behavior by saying that they did not try to elect a "pastor," but rather someone who could be strong in the midst of changes in American culture and global economics.[44] More consideration of evangelicalism, politics, and the Trump election will follow below, but what we can begin to see is that evangelicals have had a consistently pragmatic bent and a willingness to align themselves with positions or people—whether they do agree with them or not—if doing so serves ends to which they are committed. As sociologist Christian Smith argues, this constant state of being "embattled and thriving"[45] is enabled by evangelicalism's flexibility and adaptability within its pluralistic environment. Though evangelicals are sometimes seen as those who reject cultural movements, their malleability and adaptability to cultural change enable them to quickly adopt new technology and, in the case of the digital Bible, to dominate an entire industry.[46]

The Supremacy of the Bible

In the brief historical overview above, I have attempted to frame evangelicalism both as a religious tradition with a distinct set of beliefs and values and as a posture toward changes in society. I now want to narrow in on one of these values that is significant for Bible software: the Bible itself, and its status for evangelicals theologically and socially. Although scholarship seeking to define evangelicalism is relatively recent, for the last two centuries evangelicals have sought to define themselves theologically. This can take the form of declarations made on doctrinal flashpoints, such as inerrancy, open theism, or sexuality and marriage, but more often they are lists of beliefs that their authors claim all true evangelicals should adhere to or strive toward.[47] These lists can be important identity markers for some evangelicals, such as Southern Baptist Theological Seminary president Albert Mohler, who is critical of what he calls "phenomenological" accounts, including the Bebbington Quadrilateral, because they are "so vague as to be nearly useless in determining the limits of evangelicalism."[48] However, as strongly as Mohler argued for a confessional understanding of evangelicalism in his early career, he would later famously and phenomenologically flip-flop in his political views, first rejecting then later embracing Donald Trump's presidency, lending unfortunate credence to the idea that evangelicalism is often a vehicle for preserving white cultural ideals rather than theological ideas.

While one might disagree with how some evangelicals operationalize their lists of beliefs, a consistent element in each of list is an emphasis on the central importance of the Bible. For example, in the mid-nineteenth century, Anglican bishop J. C. Ryle published five distinctive principles of evangelicalism, the first of which is "the absolute supremacy it assigns to Holy Scripture," and the other four are related to sin, Jesus, and the Holy Spirit's role.[49] Eighty years later, when the NAE was founded in 1948, Harold Ockenga helped create its doctrinal statement, which consists of seven key doctrinal elements, including more historically orthodox doctrines such as a statement on the Trinity, the nature of Christ's deity, and the resurrection. Even with these additions, the NAE statement contained no reference to creeds, but instead declared its foundational belief: "We believe the Bible to be the inspired, the only infallible, authoritative Word of God."[50] In 1978, just before the creation of the Moral Majority, Canadian evangelical theologian J. I. Packer offered a sixfold definition that includes unique elements such as the importance of fellowship, but he too began with "the supremacy of Scripture."[51] John Stott, a twentieth-century British evangelical Anglican offered a threefold, Trinitarian definition of evangelicalism, but he also grounded his definition with the strong claim that "evangelical people are first and foremost Bible people, affirming the great truths of revelation, inspiration and authority. We have a higher view of Scripture than anyone else in the church."[52] Later, Packer along with theologian Thomas Oden would survey twentieth-century doctrinal lists and declarations, including those from Lausanne Movement, InterVarsity, and other evangelical groups, combining them into a longer list of sixteen emphases. Even with a relatively long list, they emphasized that "evangelical Christians, in our definition, are those who read the Bible as God's own Word, addressed personally to each of them here and now; and who live out of a personal trust in, and love for, Jesus Christ as the world's only Lord and Savior."[53]

Each of these statements confirms Bebbington's claim that "biblicism" is a key historical element of evangelicalism, and they also correspond with Warner's and Hunter's sociological analyses of evangelicalism. Evangelicals often vary in their interpretations of the Bible, but the belief that the Bible itself is in some way the "word of God" and that regular Bible engagement is an integral part of what it means to be a Christian is deeply embedded in evangelicalism, so much so that it forms an element of is cultural identity. Behind this idea is a corresponding, though largely unstated, idea that not regularly engaging scripture might be a lapsing or failure. For example, when psychologist Brian Malley asked evangelicals about their Bible reading, instead of offering straightforward responses, many felt the need to qualify their answers by saying that their reading had just recently dropped off or was not what it should be.[54] The importance of a regular practice of Bible reading is significant for Bible software, because forms of engagement such as Bible reading plans are heavily promoted in digital Bible applications. These reading plans have embedded in them a sense of completion or failure, which may be tied to one's sense of religious duty and identity.

Regularly reading one's Bible is also a common subject of social conversation among evangelicals, and its importance can be seen in the name of evangelicals' primary religious activity outside formal worship services—the "Bible study."[55] As anthropologist James Bielo has shown through extensive fieldwork, the discourse of small-group Bible studies is designed to reinforce the central importance of the Bible and being biblically based.[56] He notes that even when small-group participants differ on the interpretation of a particular passage, they always conclude that the Bible itself is a trustworthy guide for life. These discussions about the role of the Bible that take place within the Bible study setting are a significant element in developing evangelical culture, and Bible study groups continue to be an important tool for socializing newcomers in evangelical communities.[57] Sociologist Mathew Guest has shown that in church community groups that use labels other than "Bible study," the Bible still plays an important role in their social lives and in the expectations of how the group's beliefs are expressed.[58]

These claims regarding the supremacy of the Bible among evangelical leaders and the importance of Bible study among the laity leads to an important question: do evangelicals actually read the Bible regularly? According to researchers, they do. Pew reports that nearly two in three evangelicals (63%) "read scripture outside of religious services at least once a week." This is almost double the Bible reading rate of the general public (35%) and more than double that of mainline Protestant (30%), Orthodox (29%), and Roman Catholic (25%) Christians.[59] Similarly, when asked about their attitudes regarding the Bible, evangelicals report believing that the "Bible is the word of God" at much higher rates (88%) than Catholics (64%) or mainline Protestants (62%). Evangelicals are also more likely to attend a weekly Bible study (44%) than mainline Protestants (19%), Orthodox (18%), or Roman Catholics (17%).[60]

It is important to note that in the United States, these data are highly racialized. Pew classifies African American Christians as "historically black Protestant" and reports that this demographic reads the Bible at nearly the same rate (61%) as (white) "evangelicals." They also believe the Bible is God's word (85%) and attend Bible study groups (44%) at rates very close to white evangelicals. These data show that the study of American evangelicalism is inseparable from American discussion of race, and it also makes a connection with Silicon Valley, where African Americans are underrepresented across the United States' technology sector.[61] The categories of race, religion, and technology overlap directly in the Bible software industry. In my research,

I have met several minorities working in the Bible software and Christian technology spheres, but the executives and developers I interviewed for this book all presented as white.

The doctrinal lists and the public survey data above highlight the dual aspects of evangelicalism, its formal beliefs and its sociological behaviors. The Bible, then, is both a source of theological and spiritual beliefs as well as a marker of evangelicals' self-identity, sense of community, and distinction from the world around them. In the religious social shaping of technology approach, the Bible informs how evangelicals react to new technology, evaluating new tools for how they can be used with the Bible and for their relationship with God. Tayna Luhrman has argued that the printed Bible can be as a conduit for evangelical faith because it functions as a physical representation or reminder of God's presence and nearness.[62] This leads us to ask, does carrying a phone with a Bible app elicit the same sensation for evangelicals whose self-understanding is based on being a "people of the book"? To answer this question, we will now consider how evangelicals have historically approached technology.

Hopeful Attitude toward Technology

In some accounts, evangelicals are understood to be backward and culturally -resistant.[63] This is in part because evangelicalism in America is often associated with conservative stances on political issues.[64] And yet this combative stance toward the broader culture on some issues lives alongside an embrace of societal change in other areas.[65] Communications scholar Corrina Laughlin draws this theme together with evangelicalism's relationship with technology: "From the early days of radio through today, evangelicals have been early adopters of media technologies and have attempted to create a parallel culture that is the result of the twin and sometimes conflicting evangelical drives to retreat from and engage with popular culture."[66]

As Laughlin notes, evangelicals have tended to embrace media technology with a sense of hopefulness about its potential for accomplishing the mission of God. They are aware of the potential downsides of technology, but these negatives are largely framed in terms of individual morality rather than structural concerns. For example, evangelicals warn against using technology for things like viewing pornography or making an idol out of one's social media presence, but once these warnings have been given, evangelicals are drawn to the potential of media and technology both inside the walls of the church and outside, to reach those who do not yet share their faith. This orientation is characterized by the catchphrase "Technology is just a neutral tool; what matters is how we use it." By adopting a "technological frame" of hope, evangelicals are able to advocate for the aggressive implementation of new media, and this is further buoyed by what we will see in the sections below on evangelicals' embrace of entrepreneurialism and their pragmatic approach to outcomes.

Theologian and physicist Ian Barbour classifies views on technology along a spectrum of "optimism, pessimism, ambiguity."[67] I am adopting Barbour's term "optimism" and giving it a theologically inflected slant with the word "hopeful." In the nonreligious classifications above, technology itself is seen as the giver of good things, but for evangelicals, God himself has given humans technology as one of the means of producing the better world for which they hope. At its best, technology can be one of the means by which Christians can follow the words of the prophet Jeremiah, to "seek the welfare of the city" and "build houses and plant gardens" (Jeremiah 29:5–7). But evangelicals are also prone to overlook the ways in which technology can shape that mission. Media studies scholar John Ferre argues that religious people tend to think about technology in one of three ways: (1) conduit, (2) mode of knowing, and (3) social institution.[68] The term "conduit" offers a helpful analogy for the way many evangelicals approach technology because it creates the image of a tube through which one can push either morally good things or morally evil things. The most important thing, according to popular evangelical thinking, is the morality of content one puts through the tube, but the tube itself can be largely ignored. In other words, while some evangelicals take a negative stance on technology, most average users and many leaders take an instrumentalist (or conduit) view, and while some evangelical writers have offered more nuanced (ambiguous) views on technology, they ultimately advocate for its use in ministry and outreach.

This is typified in the way Billy Graham explained his view of technology: "Like most technologies, television in itself is morally neutral; it is what we do with it, or fail to do with it, that makes the difference."[69] Here we see the emphasis placed on "what we do" rather than on any inherent qualities in the technology itself, which Graham labels as neutral. This view was updated for the smartphone era in the National Association of Evangelicals' magazine, in the issue directly following the discussion on how to define "evangelical." One NAE author explicitly invokes Barbour's category of optimism: "What we personally think about changes in society is secondary to the Great Commission work of reaching our contemporary culture. . . . Let us minister from optimism."[70] Another sets up two extremes of "becoming raging technophobes railing against the nefarious uses of technology, or enthusiastic tech-evangelists who extol the virtues and possibilities of all things shiny and electronic," before arguing for a "third way." But this third way includes recognizing that "most things can be used for good or evil. Usually it all comes down to how you use it."[71]

This optimistic bent was particularly strong in the twentieth century, as evangelicals embraced microphones and cars, radio and television, and later the internet and mobile technology. As sociologist Wade Roof writes, "Evangelicalism has long used up-to-date media and information technologies in its programming and recruitment efforts."[72] Similarly, historian Randall Balmer argued that "the alacrity with which evangelicals have embraced new forms of media . . . belies the popular stereotype that they are suspicious of innovation of technology. Nothing could be further from the truth, especially in the arena of communication technology, where evangelicals have been pioneers more often than naysayers . . . . In the 20th century, evangelicals embraced electronic media with unabashed enthusiasm."[73]

Still, some evangelicals worried that this technologically powered freedom could come at a great cost. In a retelling of the formation of the National Association of Evangelicals, NAE vice president Heather Gonzales traces the formation of independent evangelical institutions, churches, mission agencies, and colleges at the beginning of the twentieth century, citing technology as a key factor in their growth. But she goes on argue that, "lacking a central organizing body, the community centered around engaging personalities and independent institutions. . . . At times they acted like rivals, weakening meaningful Christian witness."[74] It was in an attempt to bring unity among these groups while also filtering out those who might abuse the power of broadcast technology that Ockenga and others formed the NAE and other institutions like *Christianity Today* and Fuller Theological Seminary. This highlights a perpetual tension within evangelicalism: its independence from formal authority structures and entrepreneurial embrace of media sometimes enables the wrong kinds of evangelicals to gain power, prompting others to create new institutions and begin the cycle again.

Other evangelicals have expressed deeper worries, not merely about the improper use of technology, but about the nature of technology itself.

Although careful to avoid the label of Luddite or being anti-technology, they do raise structural issues about how technology can reshape society even when it is used for good. For example, in the early computer era, some evangelicals attempted to signal early warnings about how hastily adopting technology could lead to a kind of technological thinking that would get backported into churches. Kenneth Wozniak, a bank technician who also wrote on theology, argued in 1985 that Christians needed to think not just about the usefulness of a technology to a congregation, but also about the ethics relating to issues like the value of persons, dependence on technology, and themes of control.[75] Similarly, Stephen Monsma, writing from the perspective of engineering, wrestled with the idea that technology is a God-given good and yet simultaneously value-laden. In contrast to Billy Graham, he argued that technology is "not neutral."[76] As the internet emerged, some evangelicals offered critical takes on technology that urged careful discernment in everything from the use of PowerPoint to cell phones, while others drew on the work of Marshall McLuhan and Neil Postman to argue that technology might be having a stronger influence over the church's direction than most had thought.[77]

When streaming video and early forms of virtual reality became more widely available in the mid-2000s, some authors praised the new multisite model as "revolutionary,"[78] and New Testament professor Douglas Estes argued that despite the potential downsides, the "SimChurch" was a worthwhile endeavor.[79] But others, like popular Reformed blogger Tim Challies, countered that, despite his own vast reach online, face-to-face relationships should be paramount over technologically mediated ones.[80] Others contend that technology is part of what God is doing in the biblical story, but also call for careful discernment regarding technology's values and structures.[81] These discussions took on fresh significance during the 2020 pandemic, when nearly all churches worldwide had to close and use some form of internet technology to reach their congregants. While mainline and Catholic churches struggled to shift their practice and liturgy online, evangelicals had already created consumable forms church services, through podcasts, sermon downloads, and social media outlets. Their existing HEP outlook on technology made the move to fully online services easier both philosophically and practically, and it is also what enabled them to work in the digital Bible space most effectively.[82]

In the midst of these discussions that emphasize technology's goodness while debating its proper use, there have some evangelicals who display a less hopeful and more pessimistic view of technology. In his book *The Vanishing Evangelical*, former pastor Calvin Miller outlines the various causes he believes are responsible for the decline of evangelicalism, all of which are centered on an overindulgence in cultural adaptability. "Rather than speaking to our culture with a unique, prophetic voice, we have adapted to the culture and lost our vitality."[83] For Miller, the adoption of technology—including websites, Facebook, PowerPoint, and online sermons—are a part of this unbiblical adaptation to culture that ultimately undermines the evangelical message. The problem with technology is not intrinsic to technology, Miller contends, but rather the connection between technology and wider culture that he opposes. Miller's negative stance on technology, however, puts him in a small minority, and for people in the pew, their worries tend to be more practically minded. In a 2011 survey by Tyndale College in Ontario, Canada, 35% of respondents felt that church is "becoming too much about technology," and 21% of the churches reported discouraging reading the Bible electronically, but at the same time the biggest issues they faced regarding technology were inadequate budget (30.6%) and lack of volunteers (32.5%).[84] Some of the participants mentioned the intergenerational issues that often arise with technological change, but the survey did not suggest that there were more central theological issues at stake.

So while some evangelicals have taken a more negative stance on technology, as a whole, evangelicals tend to have a hopeful attitude toward technology that is consistent with their long-standing adoption of Enlightenment optimism. As Brekus writes, early evangelicals "were fervent believers in progress who dreamed of a millennial age of peace and prosperity, but they denied that progress was possible without God's grace."[85] This ties in with Pew's findings that Americans today consider technology to be the biggest improvement in their lives over the last fifty years.[86] This potent brew of American culture and evangelical optimism came together in a recent issue of *Light*, the magazine for the Southern Baptist Convention's Ethics and Religious Liberty Council, dedicated to considering technology. It contained articles addressing the economics of Bitcoin, the perils of social media, and the challenges of parenting digital natives, but its hopefulness regarding technology was also clearly evident in articles about how technology enabled faster Bible translation and how YouVersion has become "a God-ordained global movement."[87] Evangelicals, then, are not naive about the negative side of technology, and therefore do not hold a purely optimistic, conduit, or instrumentalist approach, but they draw on these views to fuel their hopefulness that God will ordain and work through their technology to accomplish his purposes. As we will see in the following section, this approach to technology is strengthened by embracing entrepreneurial business practices.

Entrepreneurialism in Business and Ministry

A second, interrelated characteristic of evangelicalism is its embrace of an entrepreneurial spirit that can be seen across a range of evangelical activity, including a willingness to experiment with models of church leadership and worship styles, and an openness to creating corporations that mix religious values with business values. Unlike Roman Catholics or High Church Protestant traditions that value preserving rituals that date back hundreds or thousands of years, evangelicals tend to reject such traditions, remain unaware of them, or simply consider them unimportant.[88] Instead, evangelical leaders are fond of saying, "The methods may change, but the message stays the same," by which they mean that while their theology is staunchly historical (they uphold the "true church"), they are free to experiment with new types of ministries, new methods of worship, and new media and technology. This phrasing takes many forms, such as "The message of the gospel is unchanging, but we may change the ways in which we present it"[89] and "While the methods may change, the message must always stay the same,"[90] but the point is the same—the *evangel* (gospel) is fixed for evangelicals (Jesus saves), but all options are on the table when it comes to sharing that message with others. For some evangelicals this includes reclaiming traditions of the past for church life when they feel it would best serve their constituents, which was exemplified in the emerging church movement.[91]

As we saw earlier, throughout their history evangelicals have been characterized by a cultural flexibility that enables them to selectively incorporate societal trends in their ministries while battling culture in other areas. In the early twentieth century, fundamentalists were known for their hostility to almost anything from the outside culture, but neoevangelicals distinguished themselves as those more open to what was happening around them. They maintained this receptivity in part by the message/method distinction, which separates theology from practice and core beliefs from their practical application. This allowed them to position themselves against those whom they perceived as abandoning doctrinal fidelity or certain forms of "traditional" morality while remaining open and flexible in the outward practice of worship, ministry, and evangelism.[92] Against this backdrop, evangelicals have tended to thrive when they employ entrepreneurial techniques to build new ministries, churches, and outreach programs that proclaim an unchanging gospel.

This embrace of the entrepreneurial spirit is pronounced in the United States, where "pulling oneself up by the bootstraps" and "living the American dream" are part of a shared cultural myth about an individual's ability to build something from nothing.[93] Though scholars like Kathryn Tanner argue that Christianity offers a counterbalance to naked capitalism, others contend that American evangelicalism is "consistent with the American ethos, [offering] a kind of spiritual upward mobility, a chance to improve your lot in the next world and also (according to the promises of some preachers) in this world as well."[94] Just as technology is considered to be neutral (or at least worth the trade-offs), evangelicals tend to accept "capitalism as not just a neutral, value-free economic system, but one that is morally good."[95] And although not all American evangelicals are Calvinistic, the connection Max Weber found between hard work and spiritual success remains deeply embedded in broader American Protestantism.[96]

This takes a special form in evangelicalism, Bielo argues, because the "born-again" transformation evangelicals experience is often extended into the secular business world, where success is interpreted as the result of following Christ.[97] Walmart's leaders, for example, employed an evangelical justification of free enterprise to control wages and thus maximize profits.[98] Putting it even more bluntly, political scientist William Connolly has argued that secular capitalism and evangelicalism bridge the sacred-secular divide through a shared ethos he calls the "evangelical-capitalist resonance machine," in which religious beliefs like the free flow of the work of God find a metaphorical resonance in capitalism's free flow of money.[99]

And yet when evangelicals discuss money (which is quite common in sermons and books), they sense a tension between their belief that material success is a sign of God's blessing and their desire to avoid falling into materialism.[100] Even so, entrepreneurs are celebrated in American society for their risk-taking, and some researchers have found that nine of out ten American business creators are affiliated with a religious tradition, most commonly probusiness evangelical churches.[101] Such business people have been a major influence on American life, and Kruse has argued that they were the primary drivers of mid-twentieth-century efforts to solidify America's religious identity through initiatives such as adding the phrase "under God" to the Pledge of Allegiance and putting "In God We Trust" on coins.[102] Similarly, in recent years a significant number of evangelicals have garnered positions of elite cultural power in not only the business world, but also among politicians, intellectuals, and artists.[103] These deep roots of entrepreneurialism in American business and religion continue to find expression both within evangelical churches and in the business people who attend them.

Scholars working in contexts outside the United States and Europe have also demonstrated connections between the Christian faith and the embrace of entrepreneurial practices. For example, business entrepreneurs from African-Caribbean ethnic backgrounds in the United Kingdom are "motivated and emboldened" by their Pentecostal faith, with some continuing to serve as bivocational pastors as they create new enterprises.[104] Likewise, private business owners in China publicly credit their success to their faith in God and use their earnings to fund evangelical churches and ministries near their churches.[105] But again this connection between faith and entrepreneurialism appears to be especially pronounced in the United States, with some scholars suggesting that one of the reasons the United States was more resistant to secularization than Europe was evangelicals' embrace of religious entrepreneurialism.[106] European churches have long been able to rely on tax revenue to fund their existence, but the American emphasis on the separation of church and state meant that US churches did have this luxury. This forced American churches to be more experimental in their approach, which in turn led to a broader variety of choices for American churchgoers.[107]

In the early 1970s, missiologist Donald McGavran made several of these strategies explicit in his book *Understanding Church Growth*, which integrates an evangelical theology of missions and evangelism with sociological research. His most well-known and most controversial argument is the "Homogeneous Unit Principle" (HUP), which claims that churches that target demographically similar people are more likely to grow.[108] Though HUP has been criticized for being functionally racist and has been largely discarded, the church growth and planting movement continues.[109] Today, among evangelicals, the "church planter"—a person who takes a risk, moving to a new city in hopes of building a thriving church from scratch—is seen in much the same light as a business entrepreneur in broader American culture.[110] To support these endeavors, evangelicals have built church-planting networks such as Acts 29, V3 Church Planting Movement, and the Association of Related Churches. Like tech incubators refining a new product, these networks help prospective church planters define a target market for their new church and then create a tailored church model that they hope will thrive among the identified demographic group.[111] The podcast "Startup," which describes itself as "A show about what it's really like to start a business," featured a story of an Acts 29 church planter, justifying the detour into the church-planting world by saying, "It's a world remarkably parallel to the tech industry, with incubators, growth metrics and, well, angel investors."[112] The series showed how the struggling startup church seamlessly blended traditional elements of an evangelical church such as preaching a conservative gospel, praying and fasting for God's guidance, and reaching out to the neighborhood, with practical touches including studying the demographics of the nearby residents, adjusting the placement of the church sign, and posting announcements on Facebook.

As we have seen, evangelicals in the business world attribute their success to their faith and pastors planting new churches blend the best of their faith with the latest information on how to build a successful business. Somewhere between creating a business and planting a church, evangelicals have also been remarkably successful at starting organizations that sell and promote products and ministries designed to appeal to believers and nonbelievers alike. Brown argues that "evangelical churches and businesses continue to mimic and use techniques found to be successful in the secular business world," including "modern advertising, promotion, and business techniques"[113] to sell music, books, apparel, and events. Some evangelicals have also begun to embrace the idea of social entrepreneurship, merging the concept of creating business for social good with church planting.[114] To help classify these different endeavors, evangelical author Andy Crouch created a fivefold taxonomy describing evangelical approaches to cultural goods: condemning, critiquing, copying, consuming, and creating.[115] Crouch argues that the first four postures have their place, but his thesis is that evangelicals should want to change culture for the better and that they will only do so through *creating* something genuinely new, beautiful, and useful. In all five of Crouch's postures, we see evangelicals managing their relationship to culture, rejecting parts of it while also maintaining their proficiency in adopting what they find useful.

This resonance between entrepreneurs in the business world who are Christians and Christian ministers who embrace entrepreneurial techniques is an important element of what has made evangelicalism one of the largest religious groups in the United States. It is also an important component of their frame regarding technology, because technology tends to be spoken of in terms of how advanced, progressive, and current it is. To successfully use technology, then, requires constant change and experimentation to which evangelicals, among all Christian groups, are uniquely suited. These traits come together in the Bible software industry, which is largely composed of evangelicals marrying their love of the Bible with their love of technology and business. This embrace of business practices can also be seen in the way they tend to judge the success of a new endeavor through the lens of pragmatism, to which we will now turn.

Pragmatism in Mission and Outcomes

Following a hopeful outlook on technology and the embrace of entrepreneurial practices, the third element that contributes to evangelicals' attraction to the digital Bible is their adoption of a pragmatic posture toward problem-solving. As a philosophy of thinking, pragmatism tends to judge the goodness of an action based on its outcome, and at its worst can fall into the justification that "the ends justify the means." But in its common form, evangelical pragmatism is an outworking of the characteristic Bebbington identified as "activism," which emphasizes the importance of missionary and social efforts. Rather than wait for a perfect solution or an ideal partner, evangelical entrepreneurialism and pragmatism work together, allowing them to join forces and experiment until something works.

This highlights two common elements of evangelicals' pragmatism: a willingness to partner with those who do not share all of their beliefs but do share common goals, and an inclination to try solutions that might work, even before all the details are figured out. Here again, the separation of message and methods comes into play, allowing evangelicals to prioritize the gospel message but be open to how it is shared—as long as it works. Examining evangelical pragmatism also requires us to acknowledge that evangelicalism is a broad movement composed of an array of different factions, reacting to each other and to the pluralistic culture around them, finding ways to work together to accomplish overlapping goals. This also indicates that all three aspects of HEP are tightly integrated and interdependent. Evangelicals' hopeful attitude toward technology is enabled not by ignoring the consequences of technology, but by a pragmatic focus on the final outcome and quick adaptation to new entrepreneurial tactics and changes in the media landscape.

This is not to say that all evangelicals entirely embrace pragmatism. In fact, some evangelicals write strongly against it, worrying that when pragmatism takes over, "theology now takes a back seat to methodology."[116] Pastor Peter Nelson laments how pragmatic thinking has come to dominate the way Christians understand the inner workings and purpose of their faith. "Pragmatism runs rampant in American Christianity. If faith does not 'work,' it lacks value." Alluding to moralistic therapeutic deism, he writes, "In recent decades, pragmatism has been recycled in the form of self-esteem doctrines, the therapeutic gospel, and the health-and-wealth message proclaimed by prosperity teachers."[117] In a similar vein, theologian Samuel Shultz critiques pragmatism on the grounds that its emphasis on efficiency leads to embracing things like printed gospel tracts that inevitably lead to a "truncated gospel," which he defines as "a pragmatic attempt to explain logically and efficiently how one becomes 'saved' and a follower of Jesus Christ the Lord."[118] He cites Charles Finney and Dwight Moody as examples of those who pioneered the simple and practical approach to evangelism that was "driven by a characteristically American spirit of idealism and resolve to 'get the job done,'" and who optimized their presentations "to get as many decisions as possible."[119]

Some Christian thinkers have also made the connection between pragmatism and essentially secular ways of being in the world. Writing in the wake of the First World War, Southern Agrarian poet Allan Tate suggested that one way of understanding secularism was as "the society that substitutes means for ends."[120] For Tate, the classical religious and natural ends of humanity have been subverted by secularization, and this process has been sped up because "the age of technology multiplies the means, in the lack of anything better to do."[121] Drawing on Tate, theologian A. J. Conyers lamented that "we have gradually lost the vocabulary and syntax necessary for speaking meaningfully, even to our own children, about nonmaterial values, nonpragmatic affections, and aims in life that exceed life itself."[122] Conyers goes on the argue that when churches focus on techniques and technology, they are emphasizing means over ends, which reduces Christianity to an essentially secular way of thinking and being, concerned more with power than love and sounding more like psychology than gospel.

And yet, for all the problems with pragmatism and utilitarianism in evangelicalism, some have cautiously touted its virtue. Noll writes, "At first glance it may seem difficult to say anything good about the American tendency to let practical and pragmatic concerns dominate over concerns of principle."[123] He critiques evangelical business leaders who operate apart from any sense of ethics and evangelists who emotionally manipulate audiences in order to claim higher numbers. But then he pivots to praise pragmatism in certain cases: "The practicality of American evangelicals which I commend . . . consists in the selective subordination of differences in principle for the purpose of accomplishing pressing tasks in practice." As examples, he cites cooperative projects including the American Great Awakening revivals, the formation of missions boards, Christian education, and moral reforms. Noll is, of course, careful to point out that each of these had its flaws, which shows that "pragmatism must not be allowed to reign without rival as Christians approach the world."[124]

In these discussions of technology and entrepreneurialism, we have already observed several examples where pragmatism played a role in evangelical decision-making. In the following section, we will look at two key examples of how pragmatism informs the way evangelicals approach problem-solving. First, the nineteenth-century missions movement serves as an example of evangelical willingness to work together with other Christians who do not align doctrinally for the common goal of evangelism. Second, evangelical political involvement can serve as a counterexample where pragmatism runs amok, and the means (political parties and candidates) undermine the ends. Although vastly different, both cases show the ways in which doctrinal and phenomenological concerns coexist within the evangelical movement.

Missions Movement

In a paper given to the Evangelical Theological Society, John Mark Hicks offers a paradigmatic defense of pragmatism from Paul's speech in the Areopagus (Acts 17), concluding that, "wherever the gospel is proclaimed, despite the reservations we might have about the pragmatism, rationalism, or rhetoric of some of the methods, let us rejoice that: Christ is preached."[125] This reasoning, while rarely so openly articulated, drove much of the Christian missions movement of the nineteenth century, where the outcome of the mission was prioritized over theological differences.

Theologian David Hilborn uses the term "pan-evangelical" to describe the broad coalitions of diverse evangelical traditions that united under the rubric of action rather than doctrine.[126] Often this action, Hilborn argues, is directed toward missionary endeavors in which many different churches and denominations cooperate toward a shared goal of bringing the gospel and Bible to people who are often referred to as "unreached people groups." However, this cooperation required that these evangelical groups self-consciously de-emphasize doctrinal differences in favor of ecumenical pragmatism. Hinson traces ecumenical pragmatism back to William Carey, a pioneering missionary to India in the late eighteenth and early nineteen centuries. He argues that Carey's influence and approach are still reflected in his Baptist successors, particularly the Southern Baptist Convention, which he cites for "the unapologetic bluntness of their pragmatism"[127] regarding the denominational groups with which they choose to associate. When Carey avoided partnerships, particularly with Roman Catholic missionaries, Hinson argues that Carey did so less because of theological differences such as baptism than because of the Baptist emphasis on volunteerism over ecclesiastical hierarchy. This meant that he favored working with groups he deemed less authoritarian in spite of theological differences.[128] In other words, his partnership choices were not principally rooted in terms of doctrinal alignment, but on how successful the partnership would be.

Carey studied the missionary work of the previous generation, including John Wesley, about whom Tomkins writes, "Whether he admitted it or not, he was a pragmatist and questions about lay sacraments, separation from the Church and women preachers were ultimately decided by what worked best."[129] Wesley also embraced the "humane use" of the technology of his day, including advanced medical care for the poor, the printing press for the transmission of his ideas and music, and modern instruments of the time.[130] This impulse to "get on with it" and figure out what works continues within Methodism and throughout evangelicalism today. Unfortunately, the motivation to work together across doctrinal lines also meant that the missions movement was often complicit in European colonialism and imperialism.[131] Sociologist Robert Wuthnow argues that this connection continues in the "globalization of American Christianity" driven by "the nation's wider participation in the international economic, political, and cultural community."[132]

The pragmatic impulse in the context of missions is also key to the evangelical push to develop Bible software applications. Sharing the gospel through the Bible is a core component of evangelicalism, and just as early missionaries embraced ships, the printing press, and other Christians with different beliefs, Bible software developers have fully immersed themselves in the technology industry, partnering with companies like Apple and Google, offering Bible translations and resources from nonevangelical denominations, all in service of the greater mission of reaching people with the Bible. For example, YouVersion has a wide variety of partnerships where it downplays theological difference in favor of a common focus on the Bible. Faithlife, too, is free to sell products and books from a variety of theological perspectives, as long the central purpose of the software is biblical interaction.

Political Engagement

If the missions movement serves as a largely positive example of evangelical pragmatism, evangelical American political engagement has not fared quite as well, especially when evangelicals align themselves with leaders and parties that appear to be at odds with their stated beliefs.[133] What began with evangelicals aligning themselves with Ronald Reagan, whose vague faith was not particularly evangelical, seems to have continued in a new form in the case of Donald Trump.[134] For some, the Trump presidency is evidence that evangelicalism is indeed merely a political force, while other evangelical leaders have argued that those who voted for Trump do not represent "true evangelicalism." Here I will continue the argument that evangelicalism is simultaneously a set of theological commitments and a posture toward the broader culture that includes the pragmatic embrace of elements evangelicals feel will help them. Sometimes it works in their favor, but as often it undermines their deepest values.

The example of Trump's election is instructive because it offers a clear case in which evangelicals adapted their views to accommodate a candidate. In 2011, the PRRI found that of all the religious groups, white evangelical had the least confidence in immoral leaders, with only 30% agreeing that immoral conduct conducted in private did not prevent governing officials from fulfilling their duties in public. But five years later, just one month before the 2016 US election, when PRRI asked the same question, the percentage of white evangelical Americans who could separate private and public conduct more than doubled from 30% to 72%. During this same period, other religious groups, including Catholics and white mainline Protestants, as well as Americans overall also reported being less likely to believe that personal immorality disqualified a candidate or elected official, but no other group was as forgiving as white evangelicals.[135] Columnist Thomas Edsall offered the following stark assessment of the shift: "What happened in the interim? The answer is obvious: the advent of Donald Trump."[136] Since Reagan's era, roughly 80% of white evangelicals have voted for Republican candidates, and Trump was no exception, earning 81% of their votes in his first campaign, appearing to confirm Edsall's thesis. So why would evangelicals, ostensibly committed to traditional morality, vote for a candidate that journalist Trip Gabriel described as "A twice-divorced candidate who has flaunted his adultery, praised Planned Parenthood and admitted to never asking for God's forgiveness?"[137] Put in terms of pragmatism, what is the end that evangelicals wanted to achieve that would cause them to change their views on a leader's morality as a means to that end?

One possible answer is American evangelicals are actually racist, sexist, homophobic, xenophobic, power-loving, pornography-using, misogynist serial divorcers, and that Donald Trump is a reflection of their true values, not a pragmatic means to an end, but someone they wholeheartedly endorse.[138] As Du Mez argues, evangelicals had long embraced a form of militant masculinity, and Trump was not an aberration, but a fulfillment of a decades-long quest for power. As a counterpoint, one could point to a handful of evangelical leaders who spoke openly against Trump, his lifestyle, and his politics or the groups of evangelicals that gathered to discuss the problems he created for their faith.[139] But the statements of an overwhelming number of prominent evangelical leaders and the history of evangelical political involvement suggests there is some truth to the connection. For example, Jerry Falwell Jr., when he was still president of the evangelical Liberty University, lent credence to the connection between Trump and evangelicals when he proclaimed, "I think evangelicals have found their dream president."[140] He even posted a picture of himself with his wife and Trump in front of one of Trump's *Playboy* magazine covers, and when asked if Trump could do anything to cause Falwell to question his presidency, he answered no.[141] Behind closed doors, he appears to have led a life that was in fact very consistent with Donald Trump's, showing that his alignment was total. But Falwell's statements were not an outlier, as dozens of other prominent evangelical leaders openly embraced him as a political, if not spiritual, ally without denouncing his immorality or racism.[142]

Rather than see this as an inconsistency, some scholars have suggested that the embrace of racist leaders is consistent with a darker thread in American evangelicalism's political engagement. Jonathan Dudley argues that, contrary to the popular belief that Falwell formed the religious Right in response to *Roe v. Wade*, "The evangelical Right, however, was not as concerned about abortion initially as it was with defending racially segregated schools."[143] Robert Jones, head of the PRRI, agrees, saying, "If you are looking for the core animating spark of the Christian-right movement, it's not abortion but private Christian universities not being able to have laws against interracial dating."[144] Carter denied Falwell's and Bob Jones's request to remain segregated, leading them to switch their allegiance to Reagan.[145] Reagan's administration initially moved to support their cases, but by 1982 they retreated in response to an outcry from civil rights groups.[146] Their association with Reagan, then, did not pay off in terms of preserving segregation, but by then they had aligned themselves with Republican economic policies and Reagan had steered the Republican Party toward evangelical priorities like overturning *Roe v. Wade*.

This long-standing connection between evangelical political allegiances and some of Trump's worst inclinations leads us to conclude that there is at least a portion of evangelicals who directly identify with his causes and embrace his racist rhetoric. But the diversity within evangelicalism also allows for a group of evangelicals who claim not to agree with Trump's personal lifestyle or all of his views, but who nevertheless voted for him. Brody and Lamb suggest that Trump intentionally supported evangelical positions he did not hold in order to gain their support.[147] Conservative commentator Hugh Hewitt argued that the evangelicals who supported Trump also did so because of their commitment to religious freedom against the onslaught of secularism, which they saw as personified in Trump's opponent, Hillary Clinton.[148]

The logic, then, was that the only way to retain the right to exercise their faith was to ensure conservative Supreme Court justices were appointed, and the only way to achieve this was to vote for a candidate who promised to appoint such judges. Unlike Falwell and Graham, who recast Trump as a moral man, these evangelicals justified their choice by saying that they knew he was immoral but that they were trying to choose a strong leader, not a "pastor."[149] This pragmatic approach bore its first fruits almost immediately when Trump appointed Neil Gorsuch to the Supreme Court. In response, evangelical pastor Robert Jeffress tweeted, "Honored to serve @POTUS on his Faith Initiative Council. He has done more in 6 mo. to protect religious liberty than any pres. in history."[150] Months after that, the Trump administration provided new exceptions to religious organizations who objected to providing birth control and abortion in their healthcare plans.[151] Trump would go on to appoint two more conservative justices before he left office in disgrace, further fueling support from some evangelicals who saw this outcome as worthy trade-off. And yet, while the vocal leaders within evangelicalism supported Trump for the issues Hewitt motioned—abortion, Supreme Court judges, and religious liberty—a survey of the broader white evangelical population showed that these issues were actually low on their list, well under the economy, national security, and immigration.[152] In other words, many evangelicals changed their moral standards, not for religious reasons that they felt were important to the common good, but because they felt he could protect their personal status and economic standing.

Regardless of the reason, however, this bargain appears to have cost evangelicals in terms of their larger stated value—sharing the gospel—with some pundits claiming to reject Christianity because of the evangelical alignment with Trump.[153] Although some of his supporters remained strong until the end, by his final year in office polls suggested that more evangelicals began to judge Trump as an ineffective leader and fewer of them supported him in the 2020 election. Still, the capital insurrection of 2021 was infused with religious symbolism, indicating an unholy alliance between faith, politics, and conspiracy theories. Although much remains to be written about the Trump presidency, his term in office may also indicate that American evangelical pragmatism has evolved into a more cynical and instrumentalist orientation toward politics than that which originated with Carter and Reagan. What began with evangelicals holding fast to a core message but remaining flexible on the methods appears to have shifted such that evangelicals are even willing to compromise on and reprioritize elements of the core message provided they see some perceived benefit, while others may have had weak Christian beliefs to begin with.[154]

This brief historical overview indicates that much of American evangelicalism is hopelessly confused about how Christian morality relates to public political engagement. But for our purposes, we can draw several important conclusions related to technology. First, American evangelicalism is not monolithic. In the missions movement, different evangelical groups worked together for the common cause of the evangelism. In the case of Trump, 25% of white evangelical voters voted against him, and of those who voted for him, they had different reasons for doing so, ranging from religious liberty and anti-Hillary votes to outright racism and economic preservation. This leads to a second point that the Trump election serves as a paradigmatic expression of evangelicalism's cultural adaptability and pragmatic approach to issues central to their faith. That is, for all the evangelicals who voted for Trump—those who openly embraced his rhetoric, those who only voted for him for religious reasons, and those who vote for him for other policy reasons—they saw him as a means to a greater end that they desired. In this sense, the Trump election has parallels to evangelical groups partnering with other churches and ministry groups who have different beliefs but who shared a common goal of missionary evangelism. However, this transaction serves a negative example because it potentially undermines other evangelical goals such as credibly sharing the gospel at home and abroad, and because in some cases, Trump was not a means but an actual desired end. And yet, this pragmatism stems from the same animating spark that allowed evangelical programmers to embrace the computer and the smartphone, despite their downsides, because of their potential to changes lives and serve their goals.

Biblically Focused, Hopeful Entrepreneurial Pragmatism

In describing the common characteristics of evangelicals in North Atlantic societies at the turn of the twentieth century, Noll outlined five basic characteristics, which summarize some of what we have observed and which point toward the way evangelicals have approached the digital Bible. He notes, first, that evangelicals "accentuated the historic Protestant attachment to Scripture," second, that they had a "shared conviction that true religion required the active experience of God," and third, that they were "extraordinarily flexible in relation to principles concerning intellection, political, social, and economic life."[155] Noll summarizes these by calling American evangelicalism a form of "culturally adaptive biblical experientialism."[156]

If Noll is correct, then the evangelical emphasis on cultural adaptability and its deep connection with the Bible combine almost perfectly in the creation of Bible software. As we will see, the hope-filled willingness to experiment and try new cultural goods like technology coupled with an interest in keeping the Bible the central object of faith are both manifest in the creation and use of digital Bibles. In addition, Noll's emphasis on the experiential nature of evangelical faith has found new meaning in the era of mobile devices, where Christian and non-Christian alike carry phones filled with apps that offer new experiences and regularly notify their users about their behavior, including their Bible reading. Together, the evangelical emphasis on the supremacy of the Bible and their HEP formed a fertile ground for the creation of the Bible software industry in the early 1980s.

4
Four Waves of Bible Software Development

In the previous chapter, we saw that evangelicals have historically sought to preserve their conservative emphasis on the Bible while also embracing modern technology, such as radio, television, and audio amplification. In this chapter, we will look more specifically into how Hopeful Entrepreneurial Pragmatism (HEP) contributed to the emergence of the commercial digital Bible industry. The printed Bible industry underwent four waves of technological advancement from the 1820s through the 1980s,[1] and we will see that the digital Bible has similarly undergone four waves of development since the 1950s: (1) preconsumer academic era, (2) desktop pastoral era, (3) popular internet era, and (4) mobile app era, with a fifth wave emerging today, the Bible as a service. The three companies that will be examined in more detail in the next chapter, Logos Bible Software, BibleGateway.com, and YouVersion's Bible app represent the second, third, and fourth wave respectively.

An examination of these waves will show several significant trends. First, while evangelicals were not active during the first wave, their commitment to the Bible and their HEP outlook on technology led them to dominate the field beginning in the 1980s with the arrival of personal home computers. Second, there was a progression in the development of Bible software from applications designed for academic study, toward desktop resources for pastoral work, and then toward online and mobile software for ordinary Christians readers. Third, the most commercially successful Bible applications were created by software developers whose evangelically oriented interest in scripture led them to develop their own Bible applications and form companies that were interconnected with evangelical institutions and ministries. Finally, as more evangelicals became involved in creating and using Bible software, the functionality and features of applications tended to reflect an evangelical outlook on the purpose and meaning of scripture.

People of the Screen. John Dyer, Oxford University Press. © Oxford University Press 2023.
DOI: 10.1093/oso/9780197636350.003.0004

The Preconsumer Academic Era, 1950s–1970s

The earliest applications of computers to biblical texts came long before computers had screens on which they could display words or keyboards that a person could use to interact with the data. In this era, which begins with the first electronic computers in the 1950s and extends through the 1970s, scholars experimented with the early computers, using them to analyze biblical texts and to produce associated materials such as concordances at much faster rates than would have been possible by hand. As the technology evolved, scholars found ways to employ computers as a part of their arguments in academic research. Neoevangelicals were not part of this initial wave of computer Bible projects, but toward the end of the 1970s, some evangelical scholars and software developers began their first forays into creating Bible software.

The Emergence of the Computer and Biblical Data Tools

Merrill Parvis at the University of Chicago pioneered the use of automation in textual criticism. In his 1952 experiment, he compared fifty manuscripts of a single verse in Luke using IBM punch cards.[2] Similarly, the Reverend John Ellison was employed by the American Philosophical society to use computers to analyze Gospel texts.[3] When the Revised Standard Version of the Bible was published in 1952, Ellison set out to create the *Complete Concordance of the Revised Standard Version of the Bible*, which would be "the first biblical concordance produced by a computer."[4] According to Deloris Burton, a chronicler of many early computational Bible projects, "Ellison, who deplored the idea that scholars with two or three doctoral degrees apiece should sit around sorting words, believed that the necessary concordance could and should be produced by a computer."[5] An article in the magazine for National Endowment for the Humanities offered this account of the magnitude of the project: "The resulting book—the first computerized concordance of the Bible—featured 300,000 entries and ran 2,157 pages."[6] After the generation of the data was complete, it took another two years to typeset the concordance. It was finally released in 1957, and popular publications praised the "massive job."[7] By today's standards, this was rather slow, but it represented an enormous leap from the thirty years it took James Strong to produce his *Exhaustive Concordance of the Bible* in 1890.

Computers for Academic Research

Through the late 1950s and 1960s, scholars continued to employ computers to build concordances, turning their sites toward other texts related to the Bible such as the Dead Sea Scrolls. As more universities purchased computers, scholars like Andrew Morton expanded on Parvis's work and created new types of research, including statistical analysis of biblical texts. Morton may be best remembered for his controversial 1963 article published in the *London Observer* called "A Computer Challenges the Church," which would lead him to be considered "one of the most prolific and long-term contributors to the field of statistical authorship determination."[8] In the article, he made the claim that one could prove through computational analysis that the apostle Paul could only have written five of the thirteen epistles that bear the name "Paul" in the introduction.[9] In the years that followed, Morton continued to perform new computational analysis, and other scholars, including Ellison himself, challenged Morton's conclusions and created new computational models to counter his claims.[10]

The computational study of the Bible continued into the 1980s under Robert Kraft, professor of early Christianity and Judaism at the University of Pennsylvania, whom Siker calls "one of the most significant pioneers in computer applications for the Bible."[11] Siker argues that Kraft is emblematic of scholars who were less concerned with interacting with the canonical text of the Bible than with "exploding the boundaries of the canonical Bible by digitizing not only the Bible but also hundreds of non-canonical writings from the same general era for historical and literary analysis."[12] Evangelicals were conspicuously absent from these early academic projects, perhaps because scholars like Morton and Kraft often argued for views of Bible that were not compatible with their own. But where the fundamentalists of the early twentieth century might have stayed away from the digital Bible, by the 1970s, some conservative scholars raised under neoevangelicalism and operating under HEP began to join in and create new digital Bible projects. Francis Ian Andersen and Dean Forbes, for example, transcribed the text of the Leningrad Codex of the Hebrew Bible and created a linguistic database of all the grammatical segments and clauses in the Old Testament.[13] In 1976, the GRAMCORD Institute was founded to create tools for computer-assisted analysis of biblical languages, and though it began at Indiana University, it was later hosted at Trinity Evangelical Divinity School by evangelical New Testament scholar D. A. Carson and eventually moved near Multnomah

Graduate School under Paul A. Miller.[14] Initially, many of the software projects developed at GRAMCORD were designed to create resources such as the *A Syntactical Concordance of the Greek New Testament*, but later GRAMCORD began to develop software that could be used by those learning Greek and Hebrew. Through the 1980s and 1990s, GRAMCORD continued to develop exegetical software representing a shift toward the end of this era from Bible software *as* scholarship toward Bible software *for* scholarship and study. And, as a project led by evangelicals, GRAMCORD also signaled a shift toward more evangelicals taking on Bible software projects.

Summary of the Preconsumer Era

For our purposes, there are two primary points of significance to observe about the work of Ellison, Morton, Kraft, and others who found ways to use computers with the Bible. First, these early adopters were not evangelicals, but instead came from a mixture of Christian traditions: Busa was a Catholic priest, Ellison an Anglican, Kraft a former evangelical.[15] Second, the nature of their work was not to interact or read the text of the Bible directly, but to use computers in service of some other end such as producing a concordance, testing a theory that could not be calculated by hand, or comparing texts. Their work was primarily for the benefit of the academy, and because computers were not yet available to purchase for home usage, their work went largely unnoticed, or at least unusable, by the common churchgoer. However, with the advent of the personal computer in the late 1970s, we will observe these trajectories shifting. Evangelicals would begin to create the majority of desktop software, and they would design it to be used by pastors and seminarians for their church work.

The Desktop Pastoral Era, 1980 Onward

While the Bible software of the 1970s had been created by academics with a particular set of research questions in mind, the new wave of applications in the 1980s were made by technologists and hobbyists, who shared overlapping attractions to emerging technology and the Bible. As Christians, they saw the Bible as a source of spiritual nourishment, and as programmers, they saw the Bible as large data set offering nearly endless opportunities to manipulate, search, and display in new ways. This experimental excitement and pragmatic approach to the emerging technology characterize many of the individual development teams in this period. This era can be divided into two stages, Infancy (1980–87) and Maturation (1987 onward), and I have chosen two presentative applications for each era that illustrate some of the key developments during that time.

Infancy: 1980–1987

The best-documented list of Bible software from this era can be found in John Hughes's book *Bits, Bytes and Biblical Studies*, published by Zondervan in 1987.[16] Giving a sense of the technology at the time, Hughes does not refer to these applications with the terms in use today, such as "Bible app" or "digital Bibles," preferring instead to highlight their primary function as "Bible concordance programs." He then subdivides these programs into those that "work with an indexed MRT (machine-readable text)" or a "nonindexed" one, a categorization that few people without specialized computer knowledge would understand.[17] This indicates that while these applications were a first step in bringing the electronic Bible out of the academy and into the church, they were not yet designed for the average churchgoer. A complete list of applications from this era can be found in the appendix to this book.

If we reflect on the locations of the companies developing these early applications, it is not surprising that several were located near the burgeoning computer hardware and software industry in California.[18] It is also noteworthy that several of the earliest companies were located in the American "Bible Belt," and three of them in Texas when Texas was becoming a major force in the marriage between evangelicals and conservative politics.[19] Below we focus on two applications, The Word Processor, the first commercial Bible software application, an example of the entrepreneurial spirit of evangelical developers, and Scripture Scanner, an example of how evangelicals began to frame Bible software as an essential tool in carrying out their mission.

The Word Processor (1982)

The first clearly documented case of a commercial Bible software application designed for nonacademic readers came from Kent Ochel and Bert Brown, who left Intel in 1980 to form Bible Research Systems (BRS) near Austin, Texas.[20] In January 1982, they released The Word Processor for the Apple IIe, claiming the mantle of "the first Bible study software on the market."[21] In this era, computer applications that interacted with a large database of information (such as the biblical text) usually required a network connection to a central computer, but BRS began "mailing out computer memory disks containing the information."[22] BRS's ingenuity attracted the attention of *Softalk Mag*, where reviewer David Hunter compared Ochel and Brown's accomplishments to Gutenberg's printing press.[23] A review in *Texas Monthly* praised the application and anticipated a hopeful day when "the industry develops hand-held computers with flat screens and letter-quality displays (all of which *will* allow you to curl up before the fire with your computer)."[24]

Ochel and Brown's work begins to show how the evangelical background of a development team could influence the Bible software industry and the consumers who used their products. One reviewer noted that the pragmatic choice to use the public domain King James Version (rather than attempting to license a more recent one) meant that The Word Processor would be better suited to Protestants than "Catholics who . . . will find that The Word Processor lacks such material as the Book of Tobit and the Book of Judith." Even so, Edwards called The Word Processor a "remarkable program" and a "monument [for] programmers."[25] Ochel and Brown continued to release updates to their flagship application called Verse Searcher through 2011, when its offices finally closed. Brown's account of his software's transition gives some sense of the scope of change in the computer industry during the application's lifetime:

> Version 1 was written for the Apple II+. As other personal computers came on the market, the software was rewritten for the TRS80 (Radio Shack), the Kaypro, the Osborne, the Commodore 64, the IBM PC, the Macintosh and all Windows computers. It was first delivered on 5.25" floppies, then 8" floppies, then 3.5" floppies, then CD, and now is directly downloaded and installed from our website.[26]

The Scripture Scanner (1984)

Another example of early Bible software whose feature set and advertising continued to signal the important role evangelicals were taking in shaping the industry was Scripture Scanner. A reviewer in *Preaching* magazine praised the addition of a word processor, writing, "The built-in word processing capability allows almost limitless possibilities for faster, more efficient research and study."[27] This distinguished it from academic and linguistically oriented applications, making it more useful for the study needs of pastors and other teachers.

Scripture Scanner also made its evangelical purposes clear in its advertising. A full-page ad in *PC Mag* ran with a headline that combined religious and technical language: "The Greatest User Friendly Story Ever Told." It went on to say: "Scripture Scanner is based on two important truths: first—The Bible is more than a book to be read; it is a *precious resource that should be used to its maximum advantage*, and second—the personal computer now gives us complete, instant access to the eternal, living, truths of God's Word."[28] Here we see the evangelical belief that the Bible is not merely an ancient book to be studied but a central source of spiritual vitality, and we can also observe both a hope-filled openness and a pragmatic embrace of new technology.

Maturation: 1988–2000

Toward the end of the 1980s, the Bible software industry began to mature alongside the burgeoning home computer market. One signal of the expanding feature set and usability of Bible software came in the 1993 publication of Jeffrey Hsu's *Computer Bible Study*. Instead of detailed technical specifications of applications found in Hughes's 1987 book *Bits, Bytes, and Biblical Studies*, Hsu wrote a chapter for each major function he felt Bible software could help pastors accomplish, such as "The Electronic Sermon," "Computer Word Study," and "Computer Topical Study." He also addressed the fledging set of online resources that provide "access to the latest theological research [and] the ability to debate and converse online with other believers and experts in religion."[29] Hsu's emphases highlight the shift from the academic work in the 1950–1970s toward software designed for use by pastors and ministry leaders, but it also reveals that Bible software was not yet created with the average Bible reader in mind.

However, by the late 1990s, popular Christian magazines such as *Christian Century* would regularly include updates about Bible software encouraging "pastors and other students of the Bible" to make use of software that was now powerful enough to help pastors keep their language skills alive and easy enough to use to introduce new aspects of scripture to lay persons.[30] A 1998 article for *Christianity Today* he cataloged the Bible software available on the market at that time, organizing it into seven categories: Academic, Professional, Devotional, Family, Add-Ons, Bible Language Tutorials, and Children's Bible.[31] The Academic category only had four applications, but there were sixteen under the Professional category and another ten under Devotional.

Several of today's largest Bible software companies were founded between 1987 and 1994, including WORDsearch (1987), PC Study Bible (1988), CDWord (1998), QuickVerse (1988), BibleWorks (1992), Logos Bible Software (1992), and Accordance Bible (1994). Not all of these applications were commercially successful enough to continue, and some, like Accordance, were focused primarily on exegetical work. In the following section, we will follow the development of two applications that were traditionally marketed to a wider audience: WORDsearch (and the products it acquired), as an example of the intertwined paths of software and evangelical institutions, and Logos Bible Software (along with the products it acquired), because it is one of the most successful combinations of business savvy with biblical studies and because it will serve as the subject of more detailed interviews in the following chapter.

WORDsearch (1987), QuickVerse (1988), the Navigators, and LifeWay

WORDsearch and QuickVerse along with the evangelical ministries Navigators and LifeWay had a long, complex, and interwoven relationship that serves as a paradigmatic example of the ways entrepreneurialism and pragmatism have driven parts of the Bible software market. WORDsearch was first released in 1987 by evangelical computer scientists James and Cheryl Sneeringer, both of whom earned a PhD in computer science from the University of North Carolina under Frederick P. Brooks. Brooks was the faculty adviser for InterVarsity Christian Fellowship, an evangelical campus ministry, through which the Sneeringers became Christians. Eventually, their shared interests in computers and Bible study led them to create the first version of WORDsearch. Two years after its release, the Sneeringers partnered with NavPress Software, the publication wing of another evangelical campus ministry, the Navigators. They later parted ways with the Navigators and bought controlling rights to NavPress Software, changing its name to iExalt and then WORDsearch.[32]

This simple back and forth with the Navigators would become increasingly complex over the next decade as the company acquired several other Bible applications and then was itself eventually acquired, further intertwining itself with other ministries and secular companies. In 2003, it merged with Epiphany Software, created by Silicon Valley programmers and engineers who were also evangelicals.[33] Epiphany had a partnership with LifeWay Christian Resources, one of the largest evangelical publishers in the United States.[34] Almost ten years later, in 2011, WORDsearch acquired another major Bible application company, QuickVerse. The story of QuickVerse, which had originally been released as Logos Bible Processor in 1998 by programmer Craig Rairdin, further demonstrates the interlinked nature of the evangelical family tree and offers an example of the ways in which business and Bible often interlocked. Rairdin partnered with a company called Parsons Technology, creators of a popular accounting software package, hoping to bundle Bible software and accounting applications and market the package to churches This represents a subtle variant of religious entrepreneurialism that attempts to serve both sacred and secular markets.

Eventually, QuickVerse was acquired by several companies, including The Learning Company (1998), Mattel (1988), Findex (2000), and finally WORDsearch in 2011. Just months after WORDsearch acquired QuickVerse, LifeWay announced that it had acquired both WORDsearch and QuickVerse, and was folding its Bible Explorer and Bible Navigator brands into a product called WORDsearch Basic, a free version of the commercial WORDsearch.[35] Then in late 2020, Faithlife acquired WORDSearch and began transitioning its customers to the makers of Logos Bible Software, which we will discuss below.[36]

The long story of InterVarsity, WORDsearch, NavPress, QuickVerse, Epiphany, Zondervan, LifeWay, and Faithlife demonstrates the deep interconnectedness of Bible software and evangelicalism's entrepreneurial spirit. WORDsearch and Epiphany were both created by software developers driven to Bible software by their evangelical convictions, while QuickVerse was started by a Christian working at a nonreligious software company that wanted to expand its presence in the church marketplace. Through a series of acquisitions and mergers, both eventually ended up at Lifeway, a Christian publishing company, and later with Faithlife, one of the largest Christian technology companies. Though evangelicalism often appears to be composed of fiercely independent churches and ministries, this story shows that these entities are often interdependent and that Bible software has been one of the key parts of evangelicalism's circuitry over the last forty years.

Logos (1991), CD Word (1989), Dallas Theological Seminary, and Zondervan

The story of Logos Bible Software began 1986 when Bob Pritchett was still in high school and he wrote first Bible software application, distributing it on bulletin boards. Years later, while working at Microsoft, he and his colleague Kiernon Reiniger were "looking for a hobby project" to create software for the recently released Windows 3.0 operating system. In December 1991, they released the first version of Logos Bible Software, which included two translations of the Bible and three additional resources. You may have noticed that the origin of Logos is similar to other applications above: two evangelicals who enjoyed technology built a business for fun around their interest in the Bible. As its story continues, we will find that it too was intertwined with other evangelical institutions, notably Dallas Theological Seminary (DTS) and its CDWord library project, which Logos eventually acquired.

CDWord, though it was ultimately a commercial failure, is another excellent example of technical innovation paired with a special regard for the Bible. At the time of CDWord's creation, commercial Bible software applications could only display English characters, not the Greek letters of the New Testament. The creators of CDWord wanted to change this and incorporate "the Greek New Testament, a full Greek lexicon, full-text Bible dictionaries and commentaries."[37] All this additional information would not fit on the floppy disks of the time, so the team determined that they would use cutting-edge CD-ROM hardware to distribute the resources.[38] Two Dallas businessmen offered to find the project and the DTS board championed it as a "major innovation in computerized Bible study."[39] CDWord was released in 1989, and reviewers called it "the largest library of reference works for Bible study on one CD-ROM"[40] and "one of the most remarkable Bible-study tools ever developed."[41]

However, while it might have been "brilliantly conceived and state-of-the-art when it was released," instructional technology professor Duane Harbin writes that "*CDWord* is a classic example of the pitfalls of the incunabular age."[42] For all its strengths, CD-ROMs proved too expensive and too difficult to configure with computers at the time, and DTS did not have the funds to update the software from Windows 2.0 to Windows 3.0 when the latter was released in 1990. This led the board to consider selling the software and the rights to the materials it had digitized to a company capable of fulfilling their original vision. J. Hampton Keathley III, a DTS student who would go on to cofound bible.org, introduced DTS and Logos, and he helped to negotiate the sales of CDWord to Logos Research Systems.[43] As with the connections between InterVarsity, WORDsearch, QuickVerse, NavPress, and Lifeway, the relationship between Dallas Theological Seminary and Logos Bible Software demonstrates that the Bible software industry is often a blend of American evangelicalism with American technological entrepreneurialism, spanning the publishing industry, higher education, and parachurch organizations.

Through the 1990s and early 2000s, Logos Research Systems underwent several transformations and rebrands as its leaders worked to find a business model that would extend their reach beyond tech-savvy pastors to a wider audience. One strategy was to redesign their system so that it could handle texts other that Bibles and Bible study, rebranding their core system as "Libronix Digital Library System."[44] Another strategy that proved more successful for Logos was building cross-platform software and creating additional services that interconnected with its core Bible software. Through the 2000s, Logos remained a Windows-only product in a crowded Bible software market that included BibleWorks, WORDsearch, and PC Study Bible, but in 2008, it became one of the first major cross-platform Bible applications when it released the first version of its platform for Apple's OS X. The following year, Logos released an app for Apple's iPhone, and in 2011, it expanded its mobile offerings to Google's Android. The company also took on non-Bible software projects including Vyrso (a reading application for popular books), Proclaim (church presentation software), and Noet (a version of its core software platform designed for those working in humanities fields). It also began publishing its own content, books, and resources under the name Lexham Press. In 2014, Logos Research Systems rebranded itself as Faithlife Corporation to position "itself for an ever-changing tech landscape"[45] and account for the wider range of products it now offered.

All of these efforts paid off, and by the early 1990s, Logos had come to dominate the PC/Windows market. As one reviewer put it: "If you run the Microsoft Windows operating environment your obvious choice is Logos."[46] The same reviewer also noted that "the *Logos* system can seem rather expensive to many, but it is hard to beat its power or versatility."[47] By version 2.0 (dubbed "Series X") and version 3.0, reviewers would continue to praise it for its ease of use, the number of resources available, and for its "innovative features, [including] automated reports, syntax searching, non-linear history, sparklines, cluster diagrams and directed acyclic graphs, to name just a few."[48]

In the following chapter, we will analyze the development team at Logos more closely, looking at its decisions-making process and how its leaders understand the relationship between faith and business, but here it is important to mention that Faithlife as a company, and its founder and former CEO Bob Pritchett in particular, exemplify HEP. The brief retelling above of the company's rise and success indicates that part of what has made Logos so dominant was its constant embrace of new technology, which allowed it to continually introduce new resources and new features, even if not all of them were initially as powerful as other platforms. This experimental approach—regularly rewriting the core software engine, expanding to new platforms, adding, dropping, and changing features—made it more agile and allowed it to reach more people than other platforms that focused on perfecting a narrower range of features and resources.[49] Part of this approach can be attributed to its founder Bob Pritchett, who, in addition to experiencing success in the Christian software market, has been recognized by nonreligious groups and publications. His ability to make pragmatic decisions that helped keep his Bible software company profitable made him the Ernst & Young Entrepreneur of the Year in 2005, put him the *Puget Sound Business Journal*'s 40 Under 40 in 2007, and kept him consistently ranked among the top twenty-five CEOs on Glassdoor.[50] Pritchett's life and work (he stepped down as CEO at the end of 2021), from his experiments with Bible software as a teen, to the various acquisitions, expansions, experiments, and even failures, embody the entrepreneurial spirit of both evangelicalism and the technology industry along with evangelicalism's core appreciation of all things Bible.

Summary of the Desktop Era

Two noteworthy trends have emerged in this analysis of the first wave of Bible software development. First, we have observed an ongoing shift in the audience of Bible software from academics in the 1970s to pastors in the 1980s to more general audiences in the 1990s. The earliest applications were largely "concordance tools," but later they added features for pastors such as built-in word processors, and over time they became more user friendly, adding more resources and visual ways of interacting with the biblical text.

Second, we have noted that the desktop Bible software industry has largely been made up of evangelical software developers with a network of connections to evangelical institutions. An important caveat to this, however, is that the most commercially successful and long-lasting applications tend to come from technologists who leave their technology professions to develop Bible software, rather than existing Bible publishing companies that attempt to enter the Bible software market. That is, individuals who seem to truly enjoy both the Bible and technology have gone on to be more financially successful than existing companies and publishers that want to expand into the technology market. Zondervan and Tyndale, for example, each released their own Bible software packages (Pradis and iLumina, respectively) but eventually folded them. In contrast, WORDsearch was created by academic technology researchers, Logos was started by former Microsoft employees, and QuickVerse began at an accounting software firm.

However, as significant as the Bible software market was at the time, its effect on regular Christian readers was limited because, in 1989, only 15% of US households owned a computer.[51] But the desktop market laid the groundwork for expanding into two emerging technologies, the internet in the 1990s and mobile devices in the 2000s. In the next two sections, we will turn our attention to these two developments, focusing on how they contributed to making the digital Bible more accessible for ordinary Bible readers and how evangelical developers took advantage of this and began to influence the way average Christians encountered the Bible.

The Popular Internet Era, 1995 Onward

As we have seen, by the 1990s, the desktop Bible software industry began to mature and reach a wider audience. But the arrival of the internet in the mid-1990s would bring major changes to society as a whole and, as evangelicals adapted to it, the creation and use of digital Bibles would change as well. As we look into this era, we will see that evangelicals' hopeful attitude toward technology extended to the internet; they tended to interpret its ability to connect with people and ideas as compatible with their values of reaching people with the gospel. The same year the internet was deregulated (1995), entrepreneurial evangelicals created a new wave of Bible websites that could display, search, and compare Bible versions without the cost of buying commercial software: Bible.com (1995), scriptures.com (1996), biblestudytools.com (1997), biblestudy.com (1998), biblegateway.com (1998), digitalbible.com (1999), ebible.com (1999), and ibible.com (1999). And even though the internet was home to all manner of content evangelicals would consider immoral or dangerous, their pragmatic lens told them it was better to bring the Bible to the internet than to miss an opportunity to share their faith. At the same time, evangelicals would find that the internet would come to influence their business and ministry models, particularly when it came to the ownership and rights of Bible translations.

In the following pages, we will trace the development of three of these websites. First, we will examine BibleGateway.com because it was one of the first websites to offer free online access to the text of the Bible, and because it has continued to be one of the most popular sites for accessing modern translations of the Bible such as the New International Version (NIV). The Bible Gateway development team will also serve as the representative from the internet wave in the following chapter. Second, we will look at Bible.com because of its historical significance as one of the first Bible-related domain names, its tumultuous business history, and its later connection to YouVersion. Finally, we will review the formation of bible.org because its founders also created a new Bible translation using a nontraditional, internet-inspired methodology.

In each case, we will observe a shift from digital Bibles being primarily a tool of academics and pastors to the something the average layperson could easily access and use. We will also continue to observe the network of evangelical institutions that contributed to this shift and the effects the new technology had on them. As previously mentioned, the creation and use of Bible websites illustrate evangelicalism's emphasis on the Bible and HEP orientation. But the internet era also brings additional nuance to the way evangelicals negotiate their relationship with society in general and technology in particular.

Bible Gateway (1993)

One of the first websites to display the text of the Bible and include search functionality was "Bible Gateway," launched December 28, 1993, by Nick Hengeveld, a student and staff member at Calvin College, with the following message board post:[52]

> Subject: [comp.infosystems.www] Anyone want to test a gateway? Body: I've got a gateway to the King James Bible put together with perl. You can find it at: http://unicks.calvin.edu/cgi-bin/bible It comes complete with a form for submitting feedback (if your browser can do forms - if it can't, just send comments or suggestions to ni...@calvin.edu).
>
> Thanks, Nick Hengeveld | RAGBRAI XXI July 25–31 | LAN Administrator | Sioux City - Dubuque | Calvin College | Team Flying Monkeys

The word "gateway" is a technical term for software or hardware that connects two computer networks. Used in this context, the "Bible Gateway" was created to connect internet users to the Bible, specifically the public domain King James Version (1611). In an article looking back at the formation and impact of Bible Gateway, Rachel Barach wrote that the site was originally intended "as an internal research tool for university students . . . [Hengeveld] wanted to employ technology to make it easy to quickly look up and read Bible text."[53] This sense of experimental discovery echoes some of the ethos of the early desktop era. However, on the current version of the "About Bible Gateway" page, it expands Hengeveld's original intent, saying that he "had a visionary passion to make the Bible digitally accessible to everyone through the very new technology at the time called the Internet."[54] This broader, more missional vision may have, in fact, been in his mind, but the original message board and the earlier, more technically oriented "About" page, along with the personal notes in Hengeveld's signature, portray a young man more interested in exploring what he could create than in what the site might eventually become. Like Bob Pritchett's high school shareware Bible software, Hengeveld's initial launch of Bible Gateway appears to be the natural outworking of someone interested in both the Bible and technology, creating what he could with the resources he had.

In 1994, the year after its initial launch, Bible Gateway expanded its access beyond Calvin College, attracting the attention of the newly forming Gospel Communications International (GCI), a nonprofit that began as Gospel Films in 1950. GCI was formed in 1995 in partnership with several other evangelical ministries, including InterVarsity Christian Fellowship, InterVarsity Press, Ligonier Ministries, The Navigators, RBC Ministries, Youth for Christ, and International Bible Society, some of which had already ventured into desktop Bible software. According to interviews with the current Bible Gateway staff, one of Hengeveld's professors had a relationship with GCI, which led to GCI purchasing Bible Gateway from Hengeveld for a "very small amount of money."[55]

Today, Hengeveld's LinkedIn page contains no reference to Bible Gateway, only mentioning that he left Gospel Films in late 1996 and currently works at Github.[56] Although Hengeveld had called it a "gateway" back in 1993, GCI did not purchase the domain name biblegateway.com until 1998, and it originally forwarded to bible.gospelcom.net. Later the branding was adjusted from "The Bible Gateway, a service of Gospelcom.Net" to "BibleGateway.com, a ministry of Gospel Communications." By December 1997, Bible

Gateway had acquired licenses to include newer, copyrighted translations of the Bible, including the NIV from Zondervan and the New American Standard Bible from Lockmann, and it was actively looking for more.[57] It had also formed a partnership with Logos Research Systems (then the name of the company that made Logos Bible Software) to provide a topical search feature, and GCI's relationship with the International Bible Society allowed it to add non-English versions as well.[58] However, reports indicated that GCI's funding began to suffer during the US financial crisis, and, in 2008, Gospel Communications decided to close its internet division and sell BibleGateway.com to Zondervan for an undisclosed amount.[59]

The sale to Zondervan highlights one of the new problems that the opportunity of the internet presented to evangelicals, namely that of funding. In an interview shortly after the sale, Zondervan CEO Moe Girkins said that BibleGateway.com was the "most widely-used Bible reference site in the world."[60] On its twenty-fifth anniversary blog post, Bible Gateway claimed to be the "Most Popular Online Bible Search Engine and Bible Reading Website" with more than fourteen billion views.[61] But for all its success, Girkins also acknowledged the challenge of providing resources that were free for consumers to read, but not free for Zondervan to host and update. As we will see in the interviews in the following chapter, the business model of ad revenue and membership is very different from the desktop Bible software model employed by Faithlife, and each model brings with it unique challenges. Desktop software requires continual sales of content and upgraded versions, while websites with advertisements need to incentivize repeated page views. In the following examples, we will see alternative strategies evangelicals employed to solve this problem with varying degrees of success.

Bible.com (1995) and YouVersion

One might expect that the most natural place for a popular Bible website would have been Bible.com. However, as we will see, not every attempt by evangelicals to merge technology and the Bible worked, and the complexities of trying to provide free ministry resources online while also maintaining an income stream sometimes created unresolvable conflicts. In 1994, Roy "Bud" Spencer Miller and his wife Betty Miller, who had previously been pastors in Texas and Arizona, felt that they could reach a wider audience through the new technology of the internet, and this led them to acquire the domain name Bible.com. On the Millers' website, they write, "Through a miraculous event, Betty and Bud were able to obtain the Bible.com website in 1994."

J. Hampton Keathley, IV (who helped broker the sale of CDWord to Logos and whom we will meet again in the story of bible.org below), recalled that in 1994 or 1995, Bible.com had been used to sell Grateful Dead T-shirts, but the owner accidentally let it expire, just as the Millers went to buy it.[62]

The Millers admit that when they bought the domain name, they had "no previous computer skills." But they said they could hear God speaking to them, telling them that they were to give up their traditional pastoring roles, and that "their ministry would continue; however, this time with a different pulpit, it would now be the sending out of their teachings and sermons through this new media . . . [The] ministry's outreach would be to the whole world." There are early indications, however, that their lack of previous technical skill resulted in a struggle to capitalize on the potential of Bible.com. Curiously, the earliest versions of the website recorded on archive.org from 1998 do not have copies of any English Bible translations.[63] Instead, the Millers wrote a series of "Bible Answers" and ran advertisements for books they had written that could were available in their online store. There was a "Bibles" page that linked to Bibles hosted on scriptures.com (registered July 1996), but the page did not include any modern English language translations, only a link to the King James Version.[64] By 2000, Bible.com stopped linking to scriptures.com and instead offered links to English translations of the Bible hosted on other websites, including the American Standard Version (ebible.org), the New English Translation (bible.org), and the New International Version (bible.gospelcom.net).[65]

This lack of a clear strategy led several investors to approach the Millers to buy the domain for as much as $100,000, but Bud and Betty refused, saying they had been "entrusted to run the site for a sacred purpose." After running the site for several years on their own, the Millers sought help in making the site more profitable through a partnership with the Buffalo-based marketing firm Tek 21.[66] However, this venture resulted in lawsuits between various entities and investors.[67] Finally in 2017, the Millers finally sold the rights to Bible.com to Life.Church, the creators of YouVersion, and the Millers moved their content to BibleResources.org.[68]

The nearly twenty-year history of Bible.com, the Millers, their investors, and eventually Life.Church and YouVersion again demonstrate that the overlap of the Bible and technology is a fertile ground for exploring the complex interactions between the evangelical desire for ministry and the demands of entrepreneurial endeavors. From their example, we can see that evangelicals in the technology industry who start a Bible project tend to see more success than nontechnical teams that attempt similar projects. But in either case, we have also observed that in the internet era, Christians struggled to balance their originating desire to "reach the world" through the "life-giving power of the Bible" with the need to generate revenue sufficient for developing and maintaining technology.

Bible.org (1995) and the NET Bible Translation

If BibleGateway.com serves as an example of a successful and influential website with humble origins, and Bible.com was a largely unsuccessful venture that finally ended up in more capable hands, bible.org is something else—an internet undertaking that attempted to radically rethink the way a Bible translation could be created and distributed. According to who.is, the domain name bible.org was registered in 1995, just two months after Bible.com, by the Biblical Studies Foundation (BSF), a Christian nonprofit organization, formed in 1994 using seed money provided by Dallas businessman T. Joe Head.[69] A version of the bible.org home page from October 1996 states that it was "dedicated to putting quality biblical research and study materials on-line."[70]

This included distinctly evangelical content, including introductions, outlines, and bibliographies for every book of the Bible, a sermon illustration database, and a "Pastoral Helps" section. Although not directory connected with Dallas Theological Seminary, much of the content of the site was written by DTS graduates and faculty, including both Keathley and his father, J. Hampton Keathley III.[71] Today, bible.org continues its evangelically oriented focus, and the current About page describes its team as "comprised of gifted Evangelical Christians." They highlight that their materials have been designed to serve an evangelistic goal because bible.org exists "to freely share the good news from God to the entire world so you can KNOW the Truth about life and eternity."[72]

But perhaps bible.org's most significant contribution was not this content, but its original translation of the Bible. After forming BSF in 1994 and registering bible.org in early 1995, its founders realized that there were no websites where one could access a modern English translation of the Bible. At this point, Hengeveld's Bible Gateway had only recently gone public and still only had access to the King James Version. According to Keathley, Head attempted to negotiate with Zondervan to license the New International Version and with Lockmann for the New American Standard Bible for use on bible.org, but those negotiations failed. In late 1995, during the annual meeting of the

Society of Biblical Literature, Head, Keathley, and several professors from Dallas Theological Seminary decided that they could create their own new translation, initially named the Internet Study Bible.[73] Eventually, it was renamed the New English Translation (NET), a double entendre referencing the fact that, according to its preface, the NET was "first Bible in history to be published electronically before it was published in print."

The NET Bible holds a unique place in translation history, with several elements setting it apart from traditional Bible translations. First, the NET Bible was released online for free as it was being written, and it invited feedback on the translation from its readers that it would incorporate into later updates. When the second beta version of the Bible was released, Daniel Wallace, New Testament editor for the NET, wrote, "The NET team has made thousands of changes to the text and notes of the New Testament since that last version."[74] Second, without the limitations of print, the creators of the NET Bible were able to create extensive study material published alongside the translation that took the form of over sixty thousand footnotes. The concept of a Bible with explanatory study notes was not new, but the NET also offered unique translator's notes explaining why a given manuscript reading was preferred or an English phrase was chosen.[75] Third, the NET Bible also distinguished itself by creating a new royalty model, allowing the text to be printed for free, but charging for the use of the notes. This "freemium" approach created a separation between the nonprofit bible.org and its for-profit partner, Biblical Studies Press, which allowed the NET Bible and bible.org to avoid many of the problems that Bible.com encountered and its for-profit press to maintain.

While these unique elements and features of the NET Bible won attention and praise, the NET Bible did not catch on as a major English translation, and at several points bible.org struggled financially.

However, in 2019, HarperCollins announced that it would license the NET Bible under its Thomas Nelson brand and launch an updated version of the text, reinvigorating the translation and further intertwining it within the world of networked evangelical entrepreneurialism.[76] Bible.org's explicit evangelical identity, its interconnection with other evangelical ministries, and its rethinking of Bible translation and distribution fit the pattern we have observed in the ventures above. But the ambitious NET translation also expands this pattern and demonstrates that when evangelicals embrace a new technology like the internet, the process can influence the way their values are expressed and understood. The internet as a technological and cultural phenomenon afforded new avenues for the creation, distribution, and use of digital Bibles that were not present in the desktop era. And yet, as evangelicals hopefully and pragmatically embraced internet technology, they found it was not simply as a neutral conduit for their message. Rather, the internet itself shaped the way they engaged with their sacred text. By inviting readers into the translation process, the NET Bible serves as an example of the social construction and social shaping of technology and of the content it produced.

Summary of the Internet Era

We have observed that the internet enabled a fresh wave of digital Bible development, which in turn enabled more people to read, search, and interact with the Bible on screen than in the desktop era. While the preconsumer era focused on academics, and the desktop era was tailored to pastors, websites like Bible Gateway brought a digitized Bible to anyone with an internet connection. In addition, the NET Bible attempted to revolutionize the way Bible translations were created and released by soliciting user feedback and incorporating it into the translation. This exposed an explicit social construction of technology process that emphasized design flexibility and made readers a part of the closure decision-making process. We have also seen that, as with the desktop era, the majority of developers and content creators were evangelicals, and many of the most successful and long-lasting ventures have ties to existing evangelical institutions and ministries. Not every venture was successful, and this was due, in part, to the way the internet brought with it the expectation that users no longer needed to pay for content. This required evangelicals to flex their entrepreneurial muscles as they experimented with new funding models that included advertising, subscription programs, and a mix of for-profit and nonprofit.

The Mobile App Era, 2000s Onward

The fourth wave of digital Bible development came in the mobile era, when electronic versions of the Bible began to be widely available on mobile devices, giving readers the power of a desktop application with the portability of a printed Bible. The start date for this era, however, is less obvious than the first electronic Bible project (1952), the first commercial desktop Bible software application (1982), or the deregulation of the internet and first Bible websites (1995). One could begin in 1998 with the release of first mobile Bible applications for the Windows CE platform, such as Laridian's PocketBible. But because the platform was largely limited to business users, one could also point to the release of the YouVersion app for the Apple iPhone a decade later in 2008. Because of this range of potential start dates and to distinguish it from the internet era beginning in 1995, I have chosen the 2000s as an approximate reference point, with YouVersion's launch as the major shift toward regular mobile Bible app usage.

As with the previous eras, evangelicals continued to be a major driving force in the development of mobile Bible apps, and we will observe an even greater shift in emphasis from biblical studies for pastors and academics toward the everyday Bible reader's spiritual growth and evangelistic outreach. Desktop Bible software had been designed for pastors and seminarians, and it emphasized features that made study and research faster and offered access to original language tools and resources. Bible websites opened some of these study features to regular Bible readers, but they could only be used when one was near a desktop computer. Just as the internet brought a different set of priorities and abilities than desktop computers, mobile devices also brought with them new functionality and a new set of values that would frame what developers could create and how users interacted with what they created.

The size and portability of mobile devices have made them one of modern culture's most prominent objects of conspicuous consumption and simultaneously made it possible to take a smartphone into a place of worship and use it in place of a printed sacred text, significantly shifting the social-material value system for Christians.[77] As Chow has argued, in the shift from print to digital, "[The] public nature of the Bible is no longer found in the physical object."[78] The digital Bible's move into sacred spaces and practices accelerated the shift in focus of digital Bible development more toward popular usage and functionality and entrenched its connections within evangelicalism. While some mobile Bible software includes study features that mimic desktop software, other popular apps like YouVersion de-emphasize study features in favor of tools like reading plans, memorization, and social connection geared toward the needs of average readers. We will see that this shift represented a multifaceted religious social shaping process as evangelicals adapted to and adopted the values of inherent in mobile technology while also working to shape it toward their own ends.

The earliest mobile Bible software application may have been developed as early as 1996 by a company called K2 Consultants, which announced that "the Holy Bible–King James Version, is now a Newton Application."[79] But while the Apple's Newton was ultimately not successful, later devices like the Palm Pilot and iPhone had significant cultural impact, and here we will trace a major application on each platform.

Palm Pilot (1996), Olive Tree (1998), and Laridian (1998)

Drew Haninger created some of the most influential of the early wave of mobile Bible apps, including BibleReader, first for the Palm platform (1998) and later for Microsoft's Pocket PC (1999). In 2000, Haninger formed Olive Tree Bible Software, which went on to produce software for Mac, Windows, iOS, Android, and Kindle Fire.[80] In yet another instance of interconnected evangelical structures, Olive Tree was later acquired by HarperCollins Christian Publishing, which also owns Bible Gateway, Zondervan, and Thomas Nelson (which later licensed the NET Bible).[81] As for the motivations of Olive Tree's founding, Haninger wrote that his own personal Christian faith was a motivation for starting Olive Tree, and he attributes its success to prayer, a focus on the Bible, and God's faithfulness.[82] After Haninger retired in 2012, Olive Tree's new CEO, Stephen Johnson, reiterated the company's desire to produce software that was not focused only on the study of the Bible, but also on the spiritual development of its users.[83] While desktop Bible software emphasized resources academics doing research or for pastors preparing sermons, Haninger and Johnson spoke of their software in evangelical terms, saying its purpose was to "change lives," enable evangelistic conversations, and deepen personal relationships with God and others.

A similar set of motivations and goals can be found in the creation of two other mobile Bible software companies, Laridian in the late 1990s and YouVersion in the late 2000s. We met Rairdin earlier as the founder of QuickVerse, but he left that company in 1998 to form Laridian, where he could shift his focus from desktop applications to mobile devices. Laridian created software for Microsoft's Windows CE platform and Palm's platform, and these applications represented a significant milestone: "In 2002, Laridian's MyBible program for Palm OS outsold all Bible software for any platform, including the most popular Bible software for Windows."[84] The means that, by 2002, mobile Bible software was outselling Logos, QuickVerse, Accordance, and all other desktop software discussed above combined. Although these numbers cannot be independently verified, if true, they would be a landmark in the ongoing shift from desktop to mobile.

These applications also represent a further shift into meeting the needs of everyday readers. Users tended to value these mobile apps less for the study features they provide, such as access to original languages, than for the availability of popular translations like the NIV. Reviewers also took notice of two other features: Bible reading plans functionality that allowed readers to track their day-to-day reading progress, something that was less feasible on a desktop application, and memorization features that allowed "readers to collect, memorize and review passages."[85] Though Laridian would eventually lose market share to iPhone- and Android-based products, the company broke ground for what would come as mobile devices grew in popularity.

The iPhone (2007) and YouVersion (2008)

By 2006, 67% of US adults owned a cellphone,[86] but only 4% owned a personal digital assistant that could run the Bible software discussed above. But a major changed occurred in 2007 when Apple launched the iPhone and Google followed with Android the next year. Smartphone ownership would rise steadily rise to 33% of US adults by 2011, to 77% by the end of 2016, and to 81% in 2019. One of the major Bible applications to ride this surge was YouVersion, produced by Oklahoma-based megachurch Life.Church (named LifeChurch.tv at the time YouVersion was first produced), and the network connections between Bible software and American evangelicalism may have reached their zenith in YouVersion.

Though YouVersion is primarily known for its Bible app today, it began in 2007 as the website YouVersion.com. The website allowed visitors to read the text of several English Bible versions, but its key innovation was allowing its users annotate verses with their thoughts and comments. YouVersion.com was envisioned as a crowdsourced, Wikipedia-like set of Bible notes contributed not by scholars though the traditional systems of publishing and commerce, but by average readers for free. It promoted the user-generated content, saying, "Discover the relevancy of the Bible through the experiences, ideas and contributions of others" and promised that users could "contribute insights and ideas through audio and video files, images, artwork and written content."[87] Like the creators of NET Bible, who invited user feedback into the process of Bible translation, YouVersion.com's creators were challenging the traditional authority structures of evangelicalism, suggesting that Bible study notes could be prepared by users, not authors, and vetted by commenters, not publishers.

At the time, teams of developers at Life.Church, led by entrepreneur Bobby Gruenewald, were experimenting with many forms of internet-based technology, including streaming Sunday morning worship and prayer services, and YouVersion.com's user-generated content was one of these attempts to embrace the openness of the internet.[88] In early 2008, Apple announced its App Store platform, where anyone could create and either sell or offer apps for free on the iPhone, and Gruenewald asked his team to create an iPhone version of the YouVersion experience for the App Store launch in July 2008. *Christianity Today* reported that in the first twenty days, YouVersion had been downloaded 183,406 times. It reached one million downloads by March 2009, and as of 2020, YouVersion remained in the top ten reference applications on App Store.

One of the founders of YouVersion, Bobby Gruenewald, often speaks of the goal of YouVersion in terms of "engagement." "What we're really trying to address is, how do we increase engagement in the Bible?"[89] The coincides with the ongoing shift that we have been observing, away from Bible software as a study platform with supplemental material and toward a focus on the text of the Bible itself. Hutchings also argues that this self-consciously evangelical identity and way of understanding scripture has led to building an application that can "train users into habits of regular Bible engagement through systems of easy access, planned routines, frequent prompts, pleasant rewards and opportunities to invest, personalise and share."[90]

But Gruenewald goes beyond that to emphasize the regularity of engagement with the Bible. Rather than include features and content like commentaries that were traditionally understood as aiding understanding, the YouVersion team made a conscious decision that their app "was not for seminary grads" but for common readers. The first blog post announcing YouVersion's Bible app mentions basic features like reading and searching, but it also emphasized its daily reading plan and the user contributions from YouVersion.com.[91] As mentioned, YouVersion would later remove the user-generated notes, not primarily because of their content, but because they did not produce the Bible engagement it wanted to see. In their place, YouVersion introduced social features designed to allow users to share Bible verses with one another and urge each other to read the Bible more often. Rather than simply making the Bible available on a mobile phone, Hutchings sees these kinds of features as a being "designed to encourage and teach users to read the Bible in particular ways."[92] This focus on the individual believer and his or her ongoing relationship to the Bible is especially suited to an evangelical understanding of the Christian faith.

YouVersion also brought a significant change in the business model of digital Bibles, and with this new model came a substantial shift in power dynamics. The first mobile apps, like BibleReader and Olive Tree, followed a revenue model similar to those behind desktop software, where users either paid a set price for the app or the app was free and users paid à la cart for content. In this model, the developers would use a portion of what they earned from their customers to pay licensing fees for the Bible translations they used. Bible websites, on the other hand, made translations freely accessible and paid for the licensing fees and bandwidth through advertisements and other subscriptions. YouVersion, however, has neither of these streams of income. It has no ads on its website or in its apps, and it does not charge users a fee to use any of its thousands of Bible versions. Instead it funds its operation entirely through millions of dollars in private donations every year.[93] Life. Church itself funded the initial stages of the app, and now it is supported by a variety of donors, one of the most prominent of whom is billionaire David Green, CEO of Hobby Lobby, who has spent much of his fortune on Bible-related products, including the Museum of the Bible.[94] This funding model changes the social shaping process in a significant way, because it potentially removes the economic feedback loop with the end user and shifts it to donors. There are, however, a new wave of Bible applications popping up all the time, ensuring that users still have a wide range of choices in the digital Bible market.

A Fifth Wave of Apps

As of 2021, a search for "bible" in both Apple and Google's stores returns YouVersion's Bible app at the top of the list, along with hundreds of additional results not included here, such as mobile versions of desktop applications like Logos and Accordance, apps for Bible websites like Bible Gateway and Blue Letter Bible, Bible apps from publishers of specific translations of the Bible, such as Crossway's ESV Bible app, and apps with trivia or Bible-related images. In addition, there is a new wave of smaller, more focused Bible apps designed for a specific group of people, such as YouVersion's Bible App for Kids, SheReadsTruth designed for female readers, and Our Bible for progressive Christians. There are also apps, like Verses, designed specifically to encourage memorization, as well as apps that prioritize listening to audio Bibles, including Dwell, which allows users to choose the voices and background music they prefer.

Ministries like the American Bible Society, the Digital Bible Society, Faith Comes by Hearing, and Crossway have also created Bible APIs (application programming interfaces) that allow programmers to access biblical texts in their apps without negotiating a complex licensing agreement. All of these innovations suggest that in addition to the four waves of digital Bible development identified here, the preconsumer era, the desktop era, the internet era, and the mobile era, a fifth era may be emerging where popular versions of the Bible become a service upon which smaller apps with personalized and customizable experiences are built.

Summary and Significance

In this chapter, I have followed the main contours of Bible software history in order to show the unique place of evangelicals in that story. We have seen that, with the exception of the preconsumer era, evangelical programmers and entrepreneurs have been quick to adapt to each new technology out of Silicon Valley, and that the most commercially successful digital Bible products—from desktop applications designed for pastors and scholars to websites and mobile apps aimed at everyday Christian readers—have been created by evangelicals who began their careers in the tech industry, while also maintaining ties to existing evangelical institutions. These connections are visualized in Figure 4.1.

These connections highlight evangelicals' unique emphasis on the importance of the Bible from the pastor's study to the congregant's pocket. Evangelical pastors want powerful research software, and evangelical readers

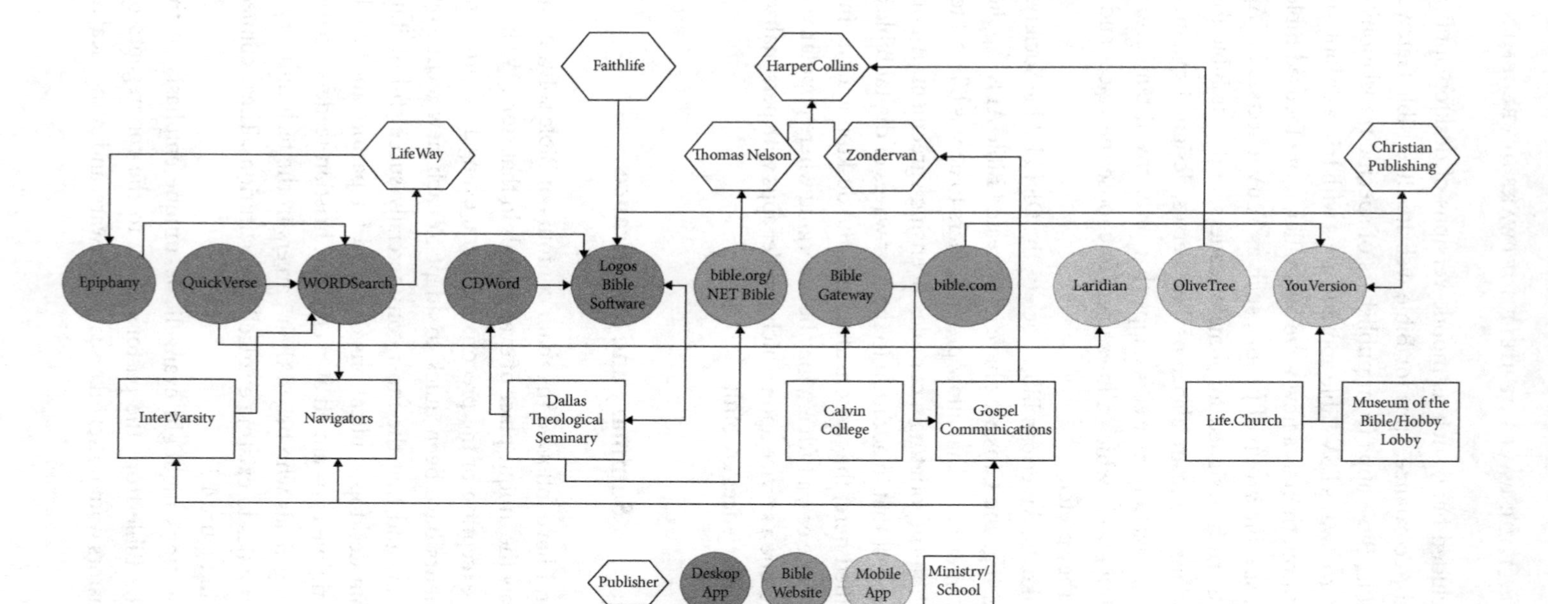

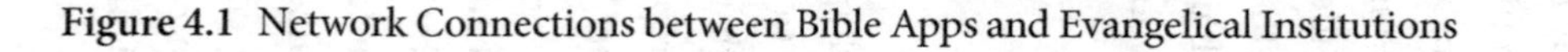

Figure 4.1 Network Connections between Bible Apps and Evangelical Institutions

want tools to help them engage with the Bible daily. Evangelical developers, powered by HEP, have been delivering apps of all kinds to meet these needs and to bring the scriptures to people around the world, some of whom do not have access to printed Bibles. In the following chapter, we will shift from this broad historical overview of Bible software toward a more intimate look at the individuals behind the apps, exploring their beliefs and motivations.

5
Programmers and the Business of Bible Software

Much of modern life runs on a layer of software that is invisible to us. Beyond the apps we see on phones and computers, the millions of lines of invisible code that run almost everything around us, including our televisions, cars, microwaves, and even kettles. If the software upon which we depend is invisible to us, even more invisible are the people who make it and their methods, motivations, and processes. In the previous two chapters, we have followed an investigative path from evangelicalism as a whole and its orientation to technological change (chapter 3) toward the narrower subject of Bible software (chapter 4). In this chapter, we will add detail to that sketch by zeroing in on the individuals actually working in the Bible software industry, men and women whom Campbell calls "religious digital creatives,"[1] before turning to look more closely at the users of the software in the following two chapters. Here we will follow three companies, each of which represents one of the waves of Bible software development, Logos Bible Software from the desktop era, Bible Gateway from the internet wave, and YouVersion from the mobile era.[2]

Drawing on focus group interviews with each of the three teams, we will explore aspects of the religious social shaping of technology from inside the companies that make the software, examining their personal "core beliefs" and "negotiation process."[3] We will also see how these values contribute to the social construction of technology (SCOT) processes, including "design flexibility" and "closure." I have organized the comments of the participants into four broad categories commonly used to understand companies: people, process, product, and profit.[4] We will begin with *people*, looking at the employees of the companies as individuals, noting their emergence into the Bible software industry from evangelical churches, institutions, and companies, as well as their open and optimistic views about the use of technology in their own religious practices.[5] Next, we will move to the *products* themselves, focusing not on a list of features in the software, but on what the developers

People of the Screen. John Dyer, Oxford University Press. © Oxford University Press 2023.
DOI: 10.1093/oso/9780197636350.003.0005

want to accomplish with those features and the overall goal they have for their applications. We will see that the language they use regarding their Bible applications is consistent with the writings and teachings of evangelicals who promote scripture engagement as a means of life change. Third, we will look at *process*, paying attention to the development models at each company as well as the internal and external inputs that contribute to the social shaping of Bible software. These include their twin identities as evangelicals and technologists and their immersion in Christian culture and the technology industry. In the final section, we will look at the *profit* models each of the three companies employs and how they shape their priorities and influence the software that makes its way onto the screens of average readers, affecting how they understand and interact with scripture. In each of these sections, we will see that the developers have deeply internalized evangelicalism's emphasis on the Bible, specifically reading it regularly for its capacity to change lives, and that their posture toward technology and business is characterized by Hopeful Entrepreneurial Pragmatism (HEP).

People: How Do Evangelicals Become Bible Software Developers?

Reflecting on my own experience, when I wrote my first bit of Bible-related software, I felt a sense of inevitability about it. I had grown up in a home where I could explore my fascination with technology through my mother's cameras and my father's computers, and I also grew up within evangelicalism, absorbing its appreciation for the Bible.[6] The more seasoned a programmer I became, the more the Bible, with its chapter-and-verse numbering system and multiple versions, looked like a database, and it was as if it were only a matter of time before I tried building something with it. As I reflected on this years later, I realized that during my first forays into Bible software, I had not first considered the relationship between technology and religion at a personal or academic level, nor had I developed a systematic view of scripture and media. Instead, I simply wanted to experience combining two of my interests to see what would happen.

In my interviews with the development teams, many of them reported following a path similar to the one I just described. That is, the combination of growing up within evangelicalism with its focus on the Bible coupled with a personal interest in technology eventually led them to Bible software, often via a web of relationships and connections. Over time, they came to see this path as guided by the hand of God, who led them to their present work. They also came to see their work as part of the larger mission of God in the world, emphasizing the importance of reaching the world with scripture using whatever tools they could imagine and create.

Immersed in Cultures of Scripture and Technology

Though the exact circumstances of how interviewees arrived at their respective companies differed, a common theme among all of them was a connection to an evangelical institution, such as a Christian college or business, along with a series of personal relationships that led them to a Bible software company. For example, of the five members at the Bible Gateway team, at least three attended private Christian universities, three had previously worked for Christian publishers or ministries, and several mentioned personal relationships that led to their employment there. Only one spoke of working for other companies first and coming to Bible Gateway through a recruiter. Similarly, several of the developers and designers at Logos Bible Software attended Christian universities or had a relational connection with Christian publishers or ministries. These connections that played a role in their eventual employment in the industry seem to be in line with Price's work showing that younger Christians often arrive at a sense of vocational calling through relationships, particularly those with mentors, and experiences within ministry contexts.[7]

It was from within this evangelical milieu of churches, universities, and ministries that their interest in technology drew them toward Bible software. One developer spoke of choosing to work at a particular Christian organization in part because "I knew somebody who had gotten a job there and they were able to help me get a job," but he also spoke of being "drawn to" the company because of its "technological progressiveness." He initially took a technical job unrelated to Bible software at a Christian publisher and later transitioned to its Bible software team before finally moving to one of the companies examined here. His journey illustrates how the overlapping identities of evangelical developers operate in concert. As an evangelical, he wanted to work for a ministry, and as a technologist, he chose the ministry based on its orientation toward technology. Another developer traced a direct connection between his dual interests in the Bible and technology when he discussed attending a Christian college where he majored in biblical studies and computer science. Upon graduation, he "went online and Googled 'bible and software,'" which led him to a job listing at the Bible software company for which he now works. Others studied at nonreligious private and state universities, but actively participated in churches or parachurch organizations that would help them maintain the strong commitment to the evangelical faith in which they had been raised. Several developers mentioned that they eventually found themselves applying their technical skill to Bible software as a personal project to "see if I could do it." In each of these cases, there was a strong connection between their evangelical upbringing, which emphasized the Bible, and their personal curiosity and openness toward new technology.

One slight alteration to the pattern of younger employees transitioning from their evangelical upbringing directly into Bible software as adults emerged in the founders of Logos Bible Software and YouVersion. Whereas Bible Gateway was created by Nick Hengeveld while he was a student at Calvin College, Logos and YouVersion were created by Christians who spent part of their early career working in nonreligious technology and business sectors. As mentioned in the previous chapter, before founding Logos Bible Software, Bob Pritchett had been an employee at Microsoft for several years. Similarly, Bobby Gruenwald and Terry Storch, cofounders of YouVersion, spent several years at technology start-ups and other ventures before joining Life.Church and later creating YouVersion. While these three men did not move directly from undergraduate work into a Bible software company, they too had been associated with evangelicalism and technology for some time before creating their respective applications. The common element, then, in all the interviewees was a background in evangelicalism with its high regard for the Bible combined with a personal interest and experience in technology. When combined, these forces eventually drew them to the Bible software industry, and, as we will see in what follows, the developers would eventually interpret their journey as a path on which God had guided them.

Guided by the Hand of God

As the interviews progressed, the discussion shifted from the initial data points of where each person went to school and the steps they took before arriving in their current position toward how they came to understand these events over time. Some scholars call this "performative belief," in which people position their actions in hindsight according to an identity they want to construct.[8] And yet as Wuthnow has argued, all religious talk is a "cultural work that people do to make sense of their lives and to orient their behavior."[9] Evangelicals in particular often look back on events in their lives and attempt to make sense of the ways in which God maybe have been directing things behind the scenes.

When considering their role as developers, none of the people I talked to used the language of "calling,"[10] as a pastor might or as is talked about in Christian living literature.[11] Instead, they often spoke in terms of falling into the profession through a series of events led by their interest in technology and their presence within evangelicalism. For example, when I asked a developer how she arrived at her present Bible software company, she laughed and said, "By accident and/or God's sovereignty—both." She went on to tell a story of how she had moved through various religious media companies, eventually leading at her present position. As she recalled each step, she recognized a complex combination of coincidences and personal choices she made along the way, but in retrospect interpreted her journey as one where God's hand was guiding her throughout. Indeed, several of the programmers mentioned that their work was spiritually significant for them and was not just ordinary software development. In talking about implementing a feature, one developer said:

> It's personal too, and that's what's nice about being here. It's not just something that I have to read off the wall and then try to implement, but it's something that's personally important to me.

Interestingly, this applied both to features that were directly Bible-oriented, such as displaying the text in a certain way or enabling a custom highlighting feature, and also to technical work that would have been largely the same when developing an application unrelated to the Bible. For example, a developer who works primarily on database maintenance viewed his job as an important work that honors God even if it is several layers away from something that might be considered directly ministry-oriented. In the last few decades, evangelicals have written extensively on the subject of a "theology of work,"[12] in which they advocate that all work, whether sacred or secular, is honoring to God. This recasts Bebbington's category of "activism," in which evangelicals express the gospel through effort in the world, to include not only explicitly religious activity such as sharing the gospel or caring for the poor, but also other forms of "secular" work, including cooking, accounting, and, in this case, coding. Though the developers did not invoke "theology of work" language explicitly, their framing of their occupation was consistent with this motif in thoughtful strands of evangelicalism.

Technology as Global Mission

In addition to seeing their overall work as honoring God, several others connected their occupation more directly to the larger mission of God in the world to reach people with the message of the gospel in the scriptures. The following comment is representative of what several individuals from all three companies indicated regarding the role of technology in reaching the ends of the earth:

> One thing that I kind of had more of a personal connection with, because it's something I had spent a lot of time working on, was I think our global reach . . . it's rewarding to think about a group of people that wouldn't be able to find scripture in their language and on the internet before, and they can now. . . . When you think about when Jesus talked about what things are going to start looking like and about the gospel going to every corner of the earth, I feel like what we do is, maybe playing a huge part of that in the twenty-first century.

What this and other interviewees seemed to be saying was that although they had not initially chosen Bible software as a vocational Christian ministry, they eventually came to see their role as a part of the greater mission of God in the world accomplished not only through traditional churches and ministries, but now through their work combining the Bible with modern technology. Some of the developers went on to express a sentiment that they viewed the entire Bible software industry as sharing a common "task" that requires multiple companies with different visions to achieve. Even as they recognized that they were competing for customers on one level, at a higher level they saw themselves as coworkers in the larger mission of the church. In closing comments, a manager at Bible Gateway summarized how the business of Bible software functioned within the larger church:

> We need a lot of different approaches and a lot of different experiences. Like I said before, there's no one prescription for scripture. There's no one UI [user interface] that makes the perfect sense for everybody in every use case. God is good because he knows we are diverse, and we have different gifts, and we need to bring them all individually to the task—to the task we're good at. Not everybody's tasked with tackling scripture engagement.

The language of "gifts" likely echoes Paul's first letter to the Corinthians, where he speaks of the "body of Christ" composed of different members or "parts," each with different "gifts" that contribute to the whole (1 Corinthians 12:12–14). Moreover, they also see modern technology, including screens and digital Bibles, as an essential part of the church's mission in the world today. A large part of that mission is "scripture engagement," which the manager above frames as the common goal of all Bible software companies. Under the category of "product" below, we will examine how the development teams use this term and how it interrelates with larger evangelical dialogues on Bible engagement. But first, we will examine their attitudes toward technology in their lives and in society.

A Range of Views on Personal Technology Usage

In popular narratives about the tech industry, those who create technology are seen as its heaviest users. However, some technology entrepreneurs, such as Steve Jobs and Bill Gates, have made headlines for limiting technology use in their homes or not allowing their children to use the products that they create.[13] Along similar lines, some researchers have begun to detect a growing trend of people rejecting newer technology like digital books and music in favor of returning to analog music and printed books.[14] So where do Bible developers fall along the spectrum of technology adoption—do they use the products they create, or do they prefer analog versions for their own personal Bible reading or in other religious situations? Put another way, to what the degree do they fully embrace a "hopeful" orientation toward technology, and how might that influence the kinds of features they create and their decision-making processes?

Much like most American evangelicals, the members of the development teams indicated that they used a combination of print and digital media, and that they determined the best media based on several factors, including what was most easily accessible, what would be the best tool for the job, and what would be most appropriate in a particular community. For example, a Bible Gateway developer juxtaposed his personal digital-only reading of the Bible with what happens in a community setting. He said that he found digital media to be more "convenient" for activities such reading books, playing video games, and communicating with others, and that this convenience was the underlying driver for moving toward engaging the Bible through "exclusively digital" means. However, he also indicated that for family devotions, he used a printed Bible. "If I sit down and do a little Bible study with my family, I research what scripture I want to go over and such online." In this case, he was not limited by an outside authority or a community expectation that said digital was inappropriate for this scenario. Rather, he had developed an internal heuristic for when print or digital was more appropriate. Members of other teams, as well as the end users we will meet in the following chapter, brought up similar boundaries at home, in Bible studies, and during church, where their own personal preference for digital Bibles was trumped by the traditions and beliefs of the larger community.

Of the three teams, members of Faithlife spoke in the strongest terms about their personal preferences for digital media and their belief that all people will eventually move away from print. One Logos team leader joked, "I'm trying to think if I know where my paper Bible is." Although several developers talked about preferring digital Bibles because of the "convenience" of always having a phone in their pockets, this leader went on to argue that he felt the electronic Bible's ease of use also made it a superior way to access truth.

> I'm a big fan of the truth and I don't always know what the truth is, but it's easier to find out digitally, it seems, than it was in paper.

The advantages of digital media were so important that he also expressed a sense of concern for those who use print media, saying, "I actually feel pity when I see people with paper books." In his view, people who use inferior printed Bibles are doing so mostly out of nostalgia and familiarity, and this might be harmful because it means they cannot get to "truth" as quickly or as easily as someone with a digital Bible. As we will see in the following chapters, this desire to find an answer and the power of digital media to help users get to that answer also surfaced in the way users spoke about their apps.

While there was a contingency of such digital-only readers, most spoke in terms of a continuum of use that was informed by the goals he or she had and what made sense within the situation. One developer explained several different scenarios and how they benefited both her personal and profession life:

> I like to mix it up. I do a little bit of both. . . . What happens with me is it depends on the day. If I've got just a few minutes between calls or something, I'll pull Bible Gateway up and do part of my reading plan and I'll do it digitally. If I have a little bit more time, if I have an hour of like solid planning and sit down and think time, I will sit down with my print Bible and do it that way.

This team leader went on to say that she intentionally uses different media in different situations in order to discover which are more helpful in a given scenario. She felt that this intentionality would enable her to perform her job better because she would understand the nuances of each media and be able to discuss their benefits and weakness with customers and friends. This discussion of different forms of media and ways of reading the Bible leads us to discuss the next important factor in the social shaping process of Bible software, the goals the development teams have for their product.

Product: What Is the Goal of Bible Software?

As I shifted from asking about the personal journeys of the individuals to discussing the ways the team as a whole understood their product and its relationship to the Bible itself, I asked questions about their goals for their software, such as, "Ideally, what do you want people to do with the software?" and, "How do you measure if your software is successful?" In almost all cases, the interviewees' answers included the term "engagement" in phrases like "Bible engagement," "scripture engagement," "engaging the Bible," and so on. In fact, this idea of "engagement" had, by the time of these interviews, become so prevalent that one team leader referred to their work as what "we in the industry buzzy-buzz world call 'scripture engagement.'" Although acknowledged in a self-deprecating way as the currently fashionable industry jargon, "engagement" nevertheless seemed to be the primary way most of the developers understood their role.

The concept of "Bible engagement" is central to both the spiritual significance and financial success of Bible software, so we will begin this section by exploring how the idea is used in the wider context of evangelicalism. This will allow see more clearly the role of Bible software in promoting and reinforcing these ideas among both evangelical readers and other users of Bible software. We will see that developers speak frequently in terms of working to increase Bible engagement using whatever means they can imagine and create, and that they enjoy the challenge of constantly trying new things to see what works. This highlights both their love of scripture and their HEP orientation to technology.

Scripture Engagement in Evangelical Thought

The term "scripture engagement" frequently appears in literature from evangelical ministries and in books written by evangelical leaders and scholars. In nearly every case, the authors express the hope that any form of Bible engagement can lead to a spiritual encounter with God that will result in a life transformation. For example, the Forum of Bible Agencies International (FOBAI) maintains the website scripture-engagement.org, where they offer this succinct definition: "Scripture engagement is encountering God's Word in life-changing ways." FOBAI also offers quotations about the Bible from writers such as Eugene Peterson, Paul Trip, and R. C. Sproul, as well as articles guiding readers on how to perform various forms of Bible engagement, including Bible reading, Bible study, Bible preaching, and memorizing the Bible. Fergus MacDonald, international director of the Taylor University Center of Scripture Engagement, offers a similar, but more developed definition:

> Scripture engagement is interaction with the biblical text in a way that provides sufficient opportunity for the text to speak for itself by the power of the Holy Spirit, enabling readers and listeners to hear the voice of God and discover for themselves the unique claim Jesus Christ is making upon them.[15]

Under MacDonald's direction, Taylor University entered a partnership with Bible Gateway to produce resources and materials on BibleGateway.

com that offer additional definitions of Bible engagement and a biblical rationale for engagement (twenty-one verses, including Psalm 1:1–3; Joshua 1:7–8; 2 Timothy 2:7; Deuteronomy 6:4–9; James 1:21–25; Luke 8:15; Colossians 3:16–17). It also guides site visitors on how they can try various forms of engagement, including praying scripture, memorizing scripture, *lectio divina*, storying scripture, and the Ignatian method. On both FOBAI's and Bible Gateway's websites, no one form of engagement is privileged over another. Rather, they are all considered valid methods that can lead to the promise of life transformation.

To underscore the importance of life change, Ovwigho and Cole from the Center of Bible Engagement present data that links increased levels of Bible engagement to decreased immoral behavior. Specifically, they argue that their data show that "the more people read or listen to the Bible, the less likely they are to engage in self-defeating behaviors such as getting drunk, abusing drugs, pornography, gambling, and destructive thoughts."[16] Other websites and articles often cite their work as proof that Bible engagement leads to positive life change. The Canadian Bible Forum summarizes these views: "The Bible informs and transforms the lives of Christians. Bible engagement matters because it sustains and nourishes faith."[17]

In evangelicalism, ideas about Bible engagement are also connected to the urge toward evangelism and sharing one's faith. The American Bible Society states, "Our mission is to make the Bible available to every person in a language and format each can understand and afford, so all people may experience its life-changing message." Here, the promise of life change is accomplished through the availability of and engagement with the Bible. Wayne Dye, a professor at the Graduate School of Applied Linguistics, Wycliff's training center for future Bible translators, wrote a series of articles showing how various cultural factors might impede or encourage scripture engagement in the context of missionary outreach.[18] For Dye, the goal of Bible engagement is spiritual healing and transformation, but the prerequisite for this outcome is that the Bible must be readily available in an understandable language to the interested reader. This is where digital Bibles play an important role. Whereas increasing the availability of scripture was once solely handled by printed Bibles, now many of these agencies see digital Bibles as a significant means of allowing more people to engage scripture and experience the life change it brings.

The Digital Bible Changes Lives

This discussion of missions, evangelism, and Bible engagement may not be at the forefront of every developer's mind as he or she does her work, but the ideas did bubble up in their discussions. Although there were some differences of nuance between the companies, one of the consistent elements was the desire to "increase" engagement in some way. One developer at Bible Gateway answered the question of what his team wanted their users to do with the software this way:

> A big part of [what we'd like people be able to do with our software] is just increasing the ability for users to engage with scripture, whatever that might look like. It's not just presenting it, but it's presenting tools that allow people to connect with it better.

Here we can observe a generally hopeful attitude toward technology along with a latent pragmatism and flexibility in that this developer considers any form of Bible engagement ("whatever that might look like") to have the possibility of being valid and beneficial. In addition, we find that the primary desired outcome of changing the software—making it "better"—is to "increase" the amount of engagement. Below, we will see that the impetus to increase engagement is also interconnected with a Bible software company's financial model and sustainability, but for now we will focus on what specifically is meant by "increasing" scripture engagement.

It appears that there are at least two distinct, but interrelated meanings of "increase." One use of the term connects back to the more evangelistic, mission-oriented idea of Bible engagement, where speaking of "increase" referred to making the Bible more broadly available to a larger number of people through the internet and internet-connected devices. A leader in the Bible Gateway team associated engagement, increase, and mission in this way: "More and more people are [accessing scripture electronically], and I think that that has increased its availability, which is part of our mission and commission." For this developer, increasing Bible engagement means using technology to increase the number of people who have access to scripture, and this goal is seen as an alignment between the mission of the company and the mission of the church in the world.

The other more common meaning of "increasing engagement" is to encourage a single individual to spend more time with the Bible him- or herself. This is clear in the way a leader of the YouVersion team stated its purpose: "Our mission statement is to help people engage in the Bible more." The key word here is "more" because the more people engage the Bible, the thinking goes, the more life change will occur. Certainly, the developers want their software to do things such as help users gain a deeper understanding of the meaning of the text or encourage them to explore cultural backgrounds behind it. But these appeared to be secondary concerns, a means to the end of the larger goal of life change. Again, this emphasis was most pronounced in the YouVersion team member, who said:

> The reality for me . . . is that the Bible transforms people's lives. If we allow our hearts to be penetrated with what God's word says, we won't be left the same. My desire is that life change happens . . . the more time that I spend in it, the more that God's going to transform your heart.

That this sounds more like the words of a pastor or Christian ministry worker is in line with the way Life.Church articulates YouVersion as fitting within its larger goals of "finding new ways to help people connect the Bible with their daily lives." Their ultimate goal was not to create YouVersion; instead, YouVersion represents only the current method in Life.Church's ongoing commitment to Bible-oriented Christian ministry. This embodies the mantra—"The methods change, but the message stays the same"—we observed in chapter 3. It also echoes Malley's argument that, within evangelicalism, "Bible reading is distinguished from other reading by its daily pattern and by the expectation that God might speak to the reader through the text."[19] The "About" page of Bible.com makes this clear:

> YouVersion represents a new frontier in Life.Church's efforts. We aren't just building a tool to impact the world using innovative technology, more importantly, we are engaging people into relationships with God as they discover the relevance the Bible has for their lives.

The team member went on to say that if something else came along that did a better job of helping people engage with scripture than YouVersion, even if that caused YouVersion to shut down, he would be pleased to know that more people were engaging with God's word through the new technology or technique. This was perhaps the most all-encompassing vision of pragmatism offered by any of the interviewees, because it included not just new technologies they could adopt or new features they could create, but any tool or method at their disposal. While not quite as comprehensive, the common point made by all the Bible software teams was that all forms of Bible engagement have the potential to help people achieve a changed life, and that life change is more important than the form or type of engagement. The technology may change, but the message of life transformation is constant.

In some sense, this mirrors what Bielo observed in his work on evangelical Bible study groups. He noticed that during Bible study meetings, the participants often differed on the interpretation of a particular passage (e.g., the group was split on whether Proverbs 13:24 encouraged or rebuked corporal punishment), but at the end of the discussion, they always returned to a social discourse on the trustworthiness of scripture.[20] Bielo argues that this pattern indicated that the goal of the study was not primarily to come to a conclusion about meaning, but to reinforce that the group's identity is centered around the shared belief that the Bible has the answers to life's questions and offers a chance to connect to the living God and to his people. Similarly, it seems that in the narratives around Bible engagement, evangelical ministries and Bible software developers are less concerned with conclusions, interpretation, or forms of engagement than with reiterating the end goal of spiritual transformation that happens through Bible engagement. In this sense, Bible engagement itself is not the end goal, but a means to the end of "life change."

A Time Machine for Pastors

The theme of individual life change through increasing Bible engagement that featured heavily in discussions with the Bible Gateway and YouVersion teams took on a slightly different nuance with the Faithlife team. While they have distinct emphases, YouVersion and Bible Gateway share several emphases that differ from those at Logos: they are both offered free of charge, they both target average Christian readers, and they both prioritize reading the Bible itself, placing all other content and functionality in a secondary, surrounding role. In contrast to this, Logos is a paid experience, designed for pastor-scholars, and encompasses much more than displaying the text of the Bible. One of the team members described it this way: "We think of our software as library software" of which the Bible is just one part.

This is consistent with the contrast Hutchings drew between the way YouVersion and Logos portray their ideal reader, "engaging with short passages of Scripture as part of their everyday routine" (YouVersion), versus a "creative scholar, hunting for new insights through attention to original languages, textual variations, historical reconstructions and libraries of commentary" (Logos).[21] These differences mean that when Faithlife developers use the terms "engagement" or "engage," they are often referring to elements of the software and other texts rather than scripture itself. For example, when referring to measuring how customers use their software, one Logos team member said:

> We have actually quite a bit of insight into whether or not the software is being used, how much is being used, how much is being engaged with, actually which parts of it are being engaged with more or less.

Notice that the term "engagement" is used here in reference to "the software," not the Bible itself. Ultimately, the Logos team sees engagement with the elements of the program (maps, lexicons, word studies, etc.) as being connected to the text of scripture, where interaction with the software and its resources implies deeper engagement with the text. However, unlike YouVersion and Bible Gateway, which see engagement primarily in terms of the individual Christian and personal spiritual enrichment, the Logos team positions the software to be used by vocational ministry professionals for the purpose of benefiting the people they minister to through writing and teaching:

> We think of Logos as being professional or consumer software, so we hope that what it's doing is that it's making an impact on people who are teaching others. So that means pastors, professors, and seminary students. It really is a time machine, and what it does is it makes easy things easier, and makes hard things possible. . . . That's why we emphasize libraries as well as just Bibles, so that we can get that professional user a very broad coverage of information very quickly to help them in their preparation.

In the discussions above with YouVersion and Bible Gateway, when the teams talked about how they wanted to make their software "better," they often spoke in terms of making it "easier" to use in hopes that doing so would "increase" engagement. In contrast, the Logos team members framed their role in making things "better" and "easier" in terms of creating tools that resulted in their users spending less time performing specific tasks. This did not necessarily mean spending less time with the software as a whole, but improving the speed, efficiency, and power of research and study-related functionality in their software. One of the company's leaders said that when he creates and refines features, he thinks about his father-in-law, who is a pastor, and how the software can reshape how he spends his time:

> I'm often thinking of him when I'm thinking about different features and thinking about how he's going to leverage it. I hope that it's just making his day-to-day job easier for preparing sermons so that when he goes and has to do other things like counseling, and whatever other things he does as a pastor, that he's not having to spend all of his time just preparing sermons. So I hope that it's saving him time to prepare sermons or study, doing a Bible study and getting that.

For YouVersion, narratives around their mission of increasing engagement meant finding ways to encourage their users to be "in the Bible more." In these and previous comments from Faithlife, we see a theme that connects Bible software engagement to the idea of "saving time" or "spending less time" on tasks around Bible study and teaching preparation. They even refer to their product as a "time machine," which hints at a fairly pragmatic approach to technology and media. Logos wants to increase Bible engagement by decreasing the amount of time a pastor spends in activities, such as pulling books from shelves and scanning them for information, that do not directly relate to preaching and teaching the Bible. Their software can do the work of scanning the library, organizing information, and prioritizing it much more quickly than a human, and theoretically, this could shave several hours off the pastor's sermon preparation time, allowing him or her to use that time for increased Bible engagement. But they also emphasized that saving time is not an end in itself, nor does it need to result in more Bible reading. Instead, the software creates a world where a pastor has more time to "do other things like counseling" and other nonteaching pastoral duties. The net result is that pastors spend less time performing the monotonous tasks of gathering resources for study, leaving them more time to spend with people. This enables pastors to contribute more to their flock's spiritual nourishment or, to put it in the terms used above, life change.

YouVersion, Bible Gateway, and Faithlife share the common goal of life change through increased Bible engagement, but here we see that this principle operates differently based on the intended audience of the software. YouVersion operates under the assumption that increased Bible engagement will almost always lead toward life change, so its people want to do whatever they can to provide the Bible on as large of a scale as possible and reach as many people as they can. As the YouVersion leader put it, "We just want to facilitate that relationship with the creator of the world and his word [and not get] in the way." In contrast, the Logos team seems less concerned with increasing the number of people reading or the number of minutes they read, than with increasing the efficiency and efficacy of the engagement. Both companies display a version of pragmatism, finding what works, but while YouVersion's is expressed in maximizing accessibility and distribution, Faithlife's professional audience leads its developers to emphasize the labor-saving functions of their software and the role it might play in enhancing a leader's overall ministry.

Experimenting with New Forms of Engagement

As we have seen, the development teams have adopted the language of "Bible engagement" to describe their primary goal. They also indicated that they continue to experiment with new features designed to encourage users to try new forms of engagement. If Logos emphasizes making the process of Bible study more efficient for pastors and YouVersion emphasizes increasing Bible engagement for everyday Christians, Bible Gateway seemed to position itself between these two, offering a more defined set of forms of engagement. Drawing on more than twenty years in operation, the team approached forms of engagement from the perspective of what they observe their users doing:

> We know from our statistics that there are three ways that people mainly engage with Bible Gateway. There is the quick hit, where they are just looking for something really fast and they go in and out; usually that's a particular verse. There is the daily devotional, which . . . they do it themselves using their own plan. And then there is the Bible study, which is an in-depth Bible study where you're looking at different passages and using all sorts of study features.

They went on to offer anecdotes about how their users go about doing each of these three primary forms of engagement (quick lookups, daily devotionals, and Bible study) and how the development team tried to make these activities easier and more functional. This involved thinking through how a longtime Christian might interact with web pages as well as how those unfamiliar with the Bible or its chapter-and-verse numbering system might need additional prompts. One of the leaders of the Bible Gateway team also referenced a partnership with Taylor University to identify additional forms of engagement beyond what users were currently doing, such as storying scripture and *lectio divina*.

They mentioned that the identification and documentation of additional forms was the first step. "The second step is finding ways to actually create functionality" that would guide users in the practices of these forms. This statement was one of the clearest indicators of how Bible software creators think through how they can use their platforms to change the behaviors of their users and encourage them to read the Bible in new ways. At the time of the interview, Bible Gateway had recently purchased Olive Tree, a mobile Bible study application, which allowed the company to offer different apps to users, each of which would be optimized for a particular form of engagement, in this case, Bible Gateway for quick lookups and reading plans and Olive Tree for study applications.[22] Similarly, in late 2014, Faithlife released a mobile application called Every Day Bible that forgoes the robust study and research features of its main Logos app in favor of a simplified interface designed to encourage daily reading. Where the desktop and mobile versions of Logos had been designed for pastors and study, the marketing materials for Every Day Bible indicate that Faithlife is targeting the average Christian reader and promoting Bible engagement that leads to life change. The initial press release reads, "Read the Bible—just a little everyday—and you'll know it better. You'll get to know God better. And you'll learn more about what it means to follow him."[23]

By creating these additional apps, Faithlife and Bible Gateway are providing software designed to encourage certain kinds of behavior with the Bible. And yet the developers continued to reiterate that no one form of engagement should be considered superior to another. They are all valid, but their effectiveness may vary from day to day and person to person. In one sense, they are attempting to influence how readers interact with the text, but at the same time, they are following the market, providing what their users are asking for. Asked what forms of engagement developers wanted from users, a team leader summed up:

> The first thing I would say . . . is that there is no one prescription for how people ought to engage with scripture. . . . There are certain signs and markers that show that a person is invested and is making this a priority in their lives. The way in which one does that can vary from person to person, and . . . even for me personally from day to day. There are so many different ways.

From this analysis, we can see that even though Logos, Bible Gateway, and YouVersion started in different eras, targeting different kinds of users with distinct products that prioritize particular forms of engagement, their developers and leaders speak in uniformly evangelical terms about Bible engagement, the need to encourage more of it, and the hope that it will result in spiritual transformation. "The more time that I spend in [the Bible]," one team leader said, "the more that God's going to transform your heart."

This viewpoint on Bible engagement and spiritual transformation is consistent with the evangelical identities we observed in the previous section, from the leadership down to the developers. They shared a common connection to evangelical churches, ministries, and institutions, and they appear to have absorbed its emphasis on the Bible, the distinction between methods and message, and the goal of transformation, leading them to a common mission of creating Bible software. In the following section, we will look more closely at the process by which they put these goals in motion and the sources from which they draw inspiration and innovation.

Process: How Is Bible Software Made?

In the print Bible industry, publishers have created and continue to create products that present the text of scripture in slightly different ways, augmenting it with images, notes, and cross-references, and wrapping it in new bindings and fresh branding. Similarly, since the advent of consumer Bible software in the 1980s, digital Bible companies have been involved in similar innovation, adding new features, new data sets, and new ways of displaying and analyzing biblical texts. By way of comparison, the print Bible industry has been operating for more than five centuries, while the digital Bible industry has existed for less than 10% of that time. However, digital technology appears to develop at a more rapid pace than mechanical print machines did, and the modern technology industry is much larger than the early print industry. This raises questions about the relationship between technology creators, technology users, the technology itself, and the cultures in which they all reside and operate.

Following the religious social shaping of technology model, we have looked at the "history and tradition" and "core beliefs and practices" of evangelicalism generally and Bible software teams more specifically. We will now turn to examine what Campbell calls the "negotiation processes" and "communal framing and discourses," but instead of looking at religious communities with formal, external authority structures, we will apply these questions to interplay between Bible software companies and their users. Faith-based organizations do not behave exactly like traditional business or churches, but there is an overlap between religious and technological practices where "technology users create rich meanings in mediated communication through their choices of media with specific symbolic features."[24] This takes on additional meaning for religious users of technology, as we have seen in developers' lives and will continue to see in users'. This interplay of the internal dynamics of faith-based software companies and the meaning making of religious technology users offers a rich example of the social shaping of technology.

Where Do Ideas Come From?

When I asked, "Where do ideas come from?" or "How do you go about creating a new feature?" the development teams mentioned the two major sources one might expect: internal ideas and external feedback. A Faithlife team member described their setup this way: "We have several project management systems where we gather ideas. And we have several different communication channels where we get ideas and feedback from users." But beyond these direct channels, they also stressed a variety of sources. A YouVersion leader responded saying, "[Ideas] happen from all over the place—I mean everywhere." A Faithlife interviewee used very similar language: "It's pretty organic. Ideas comes from everywhere."

In emphasizing "everywhere," the teams were acknowledging that software does not develop in a vacuum, but arises in part from the technological and religious environment around them. This is what the SCOT approach refers to as the *wider context* or the larger sociocultural and political milieu in which the development process takes place. For example, in the manufacturing industry, most insiders assume that new ideas come only from the manufacturing companies, but scholars have shown that innovation is distributed across users, manufacturers, and suppliers.[25] In the context of digital Bible software, this wider context includes changes in technology (from desktop to internet to mobile), as well as the larger discourse about technology and religion that takes place online and frames the expectations of what Bible software can and should do. For evangelicals, there are no definitive documents framing how one should approach technology such as those from Catholic, Anglican, and Methodist sources. Instead, the religious guiderails that frame evangelical Bible software have more to do with the kind of content they make available and developers' judgment of the success—both spiritual and financial—of the products and features they create.[26]

This leads us to explore three themes in the developers' understanding of their process: (1) internal theological instinct, where ideas come to individuals within a team or from the leadership making business-oriented decisions, (2) user feedback and analytics, where developers listen to users on message boards, in emails, and in focus groups and analyze the way people use Bible software, and (3) the technological milieu, where developers seek out—and sometimes unintentionally absorb—ideas from the technological world.

Negotiating Internal Theological Instinct and Company Needs

Both the YouVersion and Logos teams had members who had been with the companies from their origin, and both mentioned that the process of evaluating and implementing ideas had changed over time. Initially, the ideas came almost entirely from the small group of original members, but sources later grew to encompass the entire team as well as outside sources. The YouVersion representative explained the change over time this way:

> There was a day when those ideas, for the most part, were Bobby [Gruenwald's] and my ideas. The team was a lot smaller back then and looked a lot different back then, but now, honestly speaking, the team is

> laced with amazing guys, and so what I love so much right now is that a new feature launches and it's sometimes challenging to go back and figure out where the idea actually [came from].

This presents a challenge in tracing the origin of a new feature because even at the smallest level of a software development team, there are social processes that blur the source of innovation and design. Unless a product is developed by a single individual, its features will likely be created by a complex process of multiple inputs and iterations over time. However, even with this in mind, the teams shared instances when the ideas for the software had clearly originated from a lower-level developer or designer. Some of these ideas involved new methods to encourage Bible engagement, and others were related to less visible aspects of the technology. A Bible Gateway team leader, for example, gave two different examples of developers asking for permission to work on an element of the project, one based on efficiency and one based on what that person thought would be interesting:

> Some of [the ideas] come from developers, where they say, "I really need to rewrite this" or "It would be really great if we could do this."

In addition to ideas about how to optimize software and make it more stable, the development team is also responsible for creating new features and ways of interacting with the biblical text and surrounding resources. Members from each team mentioned that they gained valuable insight from listening to their users and studying their data, but they also strongly argued that, in certain cases, their development teams needed to create things that customers would never ask for. One member of Faithlife team made this quite explicit in the following statement:

> You probably heard the classic thing where Henry Ford said, "If I ask the customers what they wanted, I would have to build a faster horse [chuckles]." The customer can't look around a corner, the customer can only look at their problem. So we also make some features up basically because we can.

The phrase "because we can" embodies the HEP orientation, but in the context of the interview, this statement was not meant to be demeaning to users or to argue that the developers always know best. Instead, the team member acknowledged that a viable software enterprise needs to stay both connected to users and invested in innovation at the same time, summarizing his view as, "You definitely have to keep talking to the customer, but you also have to keep coming up with things the customer would never think of."

One area where the theological instincts of the teams came out was in the creation of social features around the biblical text. The YouVersion team leader spoke of several different attempts at creating social interactions within their app, but not because of corporate oversight or cross promotional goals. Instead, it was based on the instincts and theological belief of YouVersion's founders that engaging the Bible alongside others was an important part of a faithful Christian orientation toward scripture:

> It's not because it's Facebook . . . what drove us there originally, is that we had no data, but we just intuitively figured that if people are doing this [Bible reading] with their friends, they're going to be driven to do it more. That was kind of our intrinsic—we just believed that. I don't think that's a new concept. I actually think that's how the Bible originally started.

Here the leader wanted to make it clear that the YouVersion team was not motivated by a desire to simply mimic or copy a feature from a nonreligious application, Facebook.[27] Instead, he argued that they persisted in multiple attempts at social features, even when early versions of the feature failed to be adopted by users, because the leaders had an intuition based on their beliefs about the Bible and Christian community.[28]

For Logos, however, the social features of the app followed a different path. The team said that in early versions of the software (circa 2002) they added a "collaborative notes feature," envisioned as a place where customers "were going to have deep theological discussions and take collaborative notes in their Bibles." However, as with YouVersion's initial experiments with user-generated content, Faithlife disabled the feature because "it ended up being really a lot of people yelling at each other and then complaining about how slow it was." But over the years, as Logos sought to expand into new markets by developing additional software applications such as Proclaim and Vyrso, it eventually reorganized under the name Faithlife and created a new social networking system to integrate its products and keep customers within its ecosystem. In this case, the social functionality was driven by the larger corporate needs rather than a developer's ideas about how to make better Bible software.

This surfaces another theme, which is that as the companies around the Bible app development teams grew in size and complexity, negotiating which features and direction to pursue became more complicated. One of the team leaders at Bible Gateway spoke of balancing and prioritizing the needs expressed by parent companies Zondervan and HarperCollins. The team leader laid out the challenges of sifting through these multiple sources:

> We have a number of, if you will, I guess stakeholders in the process of product development for Bible Gateway. . . . Ideas for next iteration, next phase, next sprint, may come from the development guys. It may come from . . . from the business development guy at HarperCollins. It may come from an advertiser. It may come from a user. . . . So yeah, we start with, Where do we want to go? What do we need to accomplish? And again, that's a combination—me filtering through what do the users want? What do my corporate stakeholders want, and how do we match those together? Where do our overarching goals and objectives need to be? They vary from user growth to revenue growth to developing solutions from an infrastructure perspective that support other people in our company. So sometimes Harper is the client and we have to develop something for them.

This passage is only a small part of a larger conversation detailing how she and her team spend considerable time "filtering" and weighing inputs, requirements, and requests as part of their process for creating a road map of new features. She mentioned the two obvious sources, end users and the development team, but here she also adds inputs such as "advertisers" and "corporate stakeholders," thus expanding the scope of *relevant social groups* in the SCOT approach. Although we did not discuss any specific features that Harper or Zondervan asked Bible Gateway to create, this statement indicates that the orientation of the team toward the parent company is that of one of many "clients," including end users and advertisers. The tone of our discussion did not suggest any antagonism toward Bible Gateway's parent company, nor was there any mention of conflicts between the various business entities, only that the team had a responsibility to weigh and assess a variety of inputs in their decision-making process. In the interviewee's wording, there are times when her team "has to develop something for them [Harper]," but they were also free to determine how to go about doing so and how it might fit in with their other responsibilities.

User Feedback and Analytics

In the social shaping of technology paradigm, technological change is driven not just by producers, but also by the users of the product. This happens through direct means, such as when users offer verbal or written feedback about a product, and indirect means, when technology manufacturers observe how a product is being used, possibly in a way they did not expect, and then adapt the next iteration to meet the users' desired functionality. For example, when mobile phone makers began adding cameras, they found that users wanted to take photos of themselves, so the manufacturers added front-facing cameras to newer phones, and thus the selfie was born.[29] Similarly, the history of Twitter is one of users inventing new ways of using the service including @replies, #hashtags, and retweets, and the company responding by incorporating those features into its product.[30]

Similarly, all three development teams mentioned that users are one of their key sources for ideas on improving their product, and that they received these ideas through a variety of channels. In the 1980s, when the first commercial Bible software was introduced, companies had to rely on handwritten letters or face-to-face feedback at conventions or in focus groups. Then, in the 1990s, the internet allowed companies to use their website to gather immediate feedback on technical problems, translation issues, or feature suggestions. Some companies also created customer message boards where the company could announce new or updated products and customers could offer feedback and discuss feature requests. In the era of mobile apps like YouVersion, these feedback mechanisms are still available, but the creation of Apple's App Store and Google's Play Store introduced a new method of feedback—user reviews, which incentivizes companies to respond with updates that address user concerns. Another significant source of indirect user feedback comes in the form of analytics, where companies track user behavior, which can uncover patterns of behavior that neither the developers nor the users themselves anticipated.

The Faithlife team also mentioned that they hold regular focus groups for pastors where they ask questions, not only about what features the pastors want in Bible software, but also about the pastor's day-to-day activities.

> [We ask], "What's your problem? How can I help you solve it?" And generally what we've found is it works best if we ask them either what problems they're having or even ask them what they do, and then we come up with

ways we can improve. . . . So a classic example that would be—and a number of the times in the history of the company [it] had focused group meetings with pastors, and even just a couple pastors—[to ask], literally, "What do you do on Monday morning?" And then, "What do you do on Tuesday morning?"

The Faithlife team found that asking these kinds of questions about the duties of being a pastor often triggered an idea for a new feature that might address what the pastors talked about. The Faithlife team offered the example of a group of pastors who said that their biggest weekly task was choosing a topic for a sermon: "I have to decide what [I] am going to preach on." Before this conversation, the Logos team had only thought to create study features that would help pastors after they had selected a particular passage or topic, but this feedback prompted them to create a new tool that suggests sermon topics. A pastor can enter in a topic or a passage he or she wants to preach on, and Logos suggests several outlines and approaches to covering it in a sermon. In a second example, another group of pastors said, "I spent either Friday or Saturday looking for media to put on my slides during my sermon." The Faithlife leader said, "That was a clue to us that we should put more media integration in, so we went and picked out features that made it easier for them to find media that matched the sermon topic." Today, on the Logos startup screen, a series of tiles highlight various resources, and one space is devoted to a visual representation of a passage of scripture. It appears as if this feature was created as a subtle way to respond to the pastors' desire to more easily find media for their sermons, and also encourage other users to try the new features.

In addition to responding to direct user feedback, the era of internet and mobile apps enables software companies to log every click, tap, and scroll that users perform and use that data to discover things they might not be able to find out through direct inquiry. Analytics has come to be one of the most powerful drivers of the social shaping of technology process,[31] and all three teams gave examples of studying user data patterns as a means of making informed decisions, some of which went against the instincts or preferences of the team. In a discussion with the Bible Gateway team about their 2014 website revamp, they mentioned that much of the redesign was based on a "massive data project" that looked at factors such as the paths users took through the site, what they tended to go back to, and what terms they searched for. They also studied user preferences, offering users a choice of six different font choices for displaying biblical text. The designers chose several fonts they felt were aesthetically pleasing and technically suited for screen reading. However, after studying what users chose, they found that, much to the designers' chagrin, the "data shows the ugliest one wins."

This serves as a clear example of Bible readers shaping a Bible-reading technology, even when their preferences went against the desires of those creating the technology itself. After offering this example, the team member reflected on the dialectic of user data and developer ideas, saying, "Those kinds of things are absolutely driven by data all the time, as much as it's possible. Not that we don't have instinctive, invaluable additions to make, but we do rely as much as we possibly can on the data." In these cases, the development teams are still in control, and they have a choice as to how they are going to act upon user feedback and behavior patterns, but they have decided that it is in their best interest to adapt to their customers' preferences.

The Faithlife team also brought up the role of analytics in their business when it came to a change they made in the way the app allowed users to navigate through the biblical text. Most Bible apps and websites employ a vertical scrolling system where users swipe up or down to see the text above or below. However, most dedicated e-readers, such the Amazon Kindle and Barnes & Noble Nook, display text as pages that shift from left to right, and Logos went against almost all other Bible apps by adopting this style of moving through the text. The left-to-right paging method is often considered superior for long-form reading, but as far back as 2011, many users had taken to Logos's message board to ask for an infinite scrolling option.[32] In late 2014, the Logos team added a vertical scrolling option and announced that they had done so in direct response to user feedback.[33] One of the interviewees said, "There were people who just loved infinite scrolling and told us that was the one reason they were using YouVersion or some other product, and so we went back and added that as an option." The team also mentioned that they had been using analytics data to study why users were choosing infinite scroll over paging by studying what they did in the two reading modes. They discovered that while horizontal paging is helpful for reading, it makes it difficult to highlight a section of text that carries over onto the subsequent page. In this case, the developers had wanted to create an excellent reading experience, but users' feedback and data showed that they were more interested in an environment that favored other forms of engagement like studying, highlighting, and sharing. Again, this shows the complex shaping and negotiation process between developers' instinct and the actions of their users.

In discussing the role of analytics, the YouVersion team stressed that their primary interest is in finding ways to increase Bible engagement.

> We've got billions and billions of rows of data that we try to sift through to really help us understand how users are engaging and what are the catalytic things that if someone does this, then it is going to produce more engagement results.

As mentioned above, one of the features that YouVersion focused on was how they could encourage users to interact with their friends within the application. Their theological instinct told them social Bible engagement was essential, but the data and feedback indicated that users did not like their initial attempts. However, YouVersion indicated when the fifth version of its app was released, the data showed that users were increasingly adopting the social features. Further, the data offered specific insight into what kind of social relationships were most likely to increase Bible reading: users with at least one friend engaged with the app at higher rates than those who had no friends, and they found another step in engagement when people had seven or more friends. This led them to highlight the social features in the application. "We spend time working our best to help you, remind you, and give you on-ramps to the friends [feature] because the data tells us that if we can get you to seven or more, then the odds go way up that you're going to spend more time reading the Bible." YouVersion studied users' note contributions and verse images to see how much they were employed depending on where they were placed within the app and how that placement affected Bible engagement. Today, when a user opens YouVersion, a "Home" tab appears that highlights social features, displaying the reading habits of friends, showing a verse of the day one can share on social media, and suggesting reading plans designed to keep users coming back to the app.

Interestingly, it appears that YouVersion's user analytics show that the best way to increase and sustain Bible engagement is not to start with the Bible alone, but to start with social features and reading plans that connect to the user's larger world. One way to read YouVersion's decision is that it is somewhat nonevangelical in that it puts the Bible in second place (literally the second tab on the app). Another reading, however, would be that this is consistent with the idea that there is a social world around the Bible that undergirds evangelical culture, and that YouVersion's data show the necessity of leveraging the social side of religion as much as its textual side. At the same time, when I asked how YouVersion went about selecting the verse of the day, the team leader said, "It's probably not very theologically oriented" and explained that, initially, he combined several lists he found online. Later, he and his team adjusted the list to take into account Christian holy days throughout the year, and then began to adjust the list to include passages that were among the most highlighted and shared verses throughout their app. This mirrors the "minimum viable product" process, where developers release a product with the least amount of effort so they can immediately begin learning from user feedback and usage data, and then adapt as they continue the development process.

In some respects, this is similar to incorporating user feedback to add infinite scrolling or choose a font. However, YouVersion's verse-of-the-day list allows users to participate not merely in the shaping of the technology, but in the kind of spirituality the app emphasizes. Researchers have shown that digital Bible users tend to share Bible verses that are consistent with Christian Smith's moralist therapeutic deism (MTD), such as Philippians 4:13, rather than theologically oriented passages like John 3:16.[34] It is possible that YouVersion's choice to highlight popular verses is forming a feedback loop that privileges MTD passages that, in turn, further reinforce MTD viewpoints. Some evidence of this can be found on Bible Gateway's blog, which, in 2009, reported that the most popular verse was John 3:16, but by 2018 it had fallen to second place, and Jeremiah 29:11 became the most popular verse on the website.[35] These shifts may be viewed as both a reflection of and a result of evangelicalism's tendency to emphasize the individual's personal faith, as well as its tendency to lean into "what works" over what might be more theologically grounded. This also hints at ways in which the larger technological environment, including social media, can shape spirituality and the development of Bible apps.

Technological Milieu and Research

In the examples above, app features came about from the interplay between the theological instincts of the software development team and the feedback and data from users. But it is also important to recognize that both of these groups are immersed in modern technological culture, including the world of hardware, software, and innovation, or what the SCOT approach calls the *wider context*. Developers indicated that important sources of inspiration came from other applications and conventions outside the Bible software industry. One of the senior leaders at Faithlife made this explicit:

> I actually spend time looking at what other people are doing, looking at what's happening in research, in information retrieval, visualization. We have, I'm embarrassed to say at times, looked at something someone else has done in another field and said, "That just looks so awesome, is there a way we could apply it to our field?"

He offered additional examples of researching new software techniques that might be applicable to Bible software, from taxonomies in academic fields such as anthropology to visualizations used by news media organizations. Where YouVersion's creation of social features highlights an evangelical theological instinct that prioritizes community and engagement, Logos's study of other software demonstrates the willingness of evangelicals to optimistically seek out new technology that they can absorb for their own purposes. This demonstrates Faithlife's hopeful attitude, which encourages the company to seek out technology that looks "awesome" and "cool," attempt to build something with it, and then see if it works.

Another example of the way the larger context of technological development can shape Bible software came in YouVersion's move from a "hamburger" menu system to a tab-based menu system. When Apple started allowing applications in the App Store in 2008, a standard navigation paradigm was to have five buttons in a row at the bottom of the screen. Over time, there was a shift to a "hamburger" menu, three stacked horizontal lines (≡) in the upper right or upper left of an application that allowed more than five functions to be displayed. But after several years of experimentation, the design community concluded that the hamburger menu made navigation more difficult and hid the application's core functionality.[36]

The YouVersion team leader mentioned that the development and design team (i.e., not the leadership team) had been keeping track of these trends in popular apps. The leader recalled, "Our designers have hated the hamburger menu for years—hated it." But the management team was reluctant to include the menu change on the product road map because they had other features in mind and changing the menu might interfere with those priorities. As the leader put it, "My concern [was that] we've navigated projects like this in the past where it'll derail the entire direction of the app for a year. . . . We've got a long road map of things that we *know* will help Bible engagement." When the time came to decide whether to put in the work it would take to move from the hamburger to a tabs interface, the senior leader took a pragmatic approach, telling his team, "Help me understand the ROI [return on investment]. . . . How are you asking me to shut all of that down to go make the app prettier?" They responded with research articles showing that tabbed menus led to "increased engagement" in other applications. That key phrase—"increased engagement"—along with research to back it up was enough to convince the YouVersion leadership that it was worth the additional time to make the change.

In this example, we see all three of the above categories—ideas from the development team, feedback from users, and the wider context of software techniques and methodologies—interacting to produce a change in the app. This example demonstrates that digital Bible features arise from an interwoven tapestry of technology, theology, experimental innovation, and, as we will see below, financial needs. No one source appears to govern the direction of any of the companies. Instead, they go through a process of filtering and prioritizing competing desires and needs, sometimes favoring their instincts and sometimes leaning toward more rigorous research and data analysis. In the words of the YouVersion leader, "How do we just see what God is blessing and what he's on and just get behind it?" And yet to do God's work in Bible software requires sources of capital, and in the next section we will consider how the business model of each company affects its process and product.

Profit: How Does Bible Software Sustain Itself?

Thus far, we have seen that the software teams are deeply evangelical, interested in promoting Bible engagement for the purpose of changing lives, and inspired by a variety of internal and external sources. In this final section, we will examine how these factors interact by looking at three themes in the way the companies define success and its connection to their financial models. First, evangelical Bible developers show flexibility and ingenuity in their varied financial models. Second, they share a common goal of increasing Bible engagement that leads to both spiritual growth and financial stability. Third, they are aware that Bible engagement, missional outreach, and financial profit are deeply interwoven, and they see it as a part of their mission to the world.

Flexible Financial Models

As we noted in chapter 3, scholars have explored the complex relationship between Christianity and financial success, or more specifically, evangelicalism and capitalism. Hayes, building on the work of Max Weber, traces how evangelicalism arose alongside American capitalistic enterprises and came to embrace and incorporate their ideals through the late nineteenth and early twentieth centuries.[37] This can be seen in a recursive loop where increasing Bible engagement leads to spiritual growth and also to higher profits, which in turn allows for more Bible reading and more spiritual and financial growth. This echoes Connolly's concept of the "evangelical-capitalist resonance machine," discussed in chapter 3, in which evangelical values and capitalism reverberate and reinforce one another.[38]

Each of the Bible companies here began as free tool without a clear financial model, and the business savvy and flexibility of the HEP approach can be seen in the way they pivoted to a sustainable business. Bob Pritchett, founder of Faithlife, the parent company of Logos, created his first Bible software application in high school for fun and offered early versions of Logos as shareware (that is, payment was a "tip" rather than required). Bible Gateway also began as a free (and ad-free) experience on Calvin College's web servers. YouVersion, too, was an experimental website from within Life.Church's large technical division.

Over time, however, each of the Bible applications had to find a revenue stream that could pay for employees, servers, Bible licenses, and other expenses. For Logos, which began just before the internet went public, the business model became straightforward software and content sales. The company is sustained both by the initial purchase of its software library system and by purchases of books, resources, and subscriptions to content. Bible Gateway, however, began as a free-to-access website. As one of the first Bible websites, its popularity grew quickly, and with that popularity came rising hosting costs. This led its first parent company, Gospel Communications, to place ads for Christian products around displayed biblical texts. Today, ads remain one of its primary revenue sources.[39] YouVersion has an entirely different financial model based on relationships with publishers and support from major donors. It was initially funded by Life.Church and is now supported by a variety of donors. As mentioned in chapters 1 and 4, a prominent donor is billionaire David Green, founder of Hobby Lobby, who has spent much of his fortune on Bible-related ventures, including the Museum of the Bible and YouVersion.[40] Today, the Bible app displays a donation function to regular users, inviting them to "support what God is doing through YouVersion, and help build His Church by investing in His Kingdom."

Relational Bible Engagement

So what do these different systems—sales (Logos), advertisements (Bible Gateway), and donations (YouVersion)—have in common, and how do they influence the features of the applications? As we have seen above, the companies have a shared belief that the Bible is divinely inspired and that engagement with it, through any form or any technology, is good for spiritual growth. Unlike printed Bibles, software creates an ongoing, interactive relationship with the text. As we will see, increasing engagement on software platforms results not only in life change, but in financial stability. For Logos, increasing the number of people who engage with their software or who buy additional products results in additional revenue. This ensures the company's long-term sustainability, which in turn allows users to continue using their product and engaging scripture. The longer users stay on the Bible Gateway site and the more content they view, the more ads they see and the more revenue the company can bring in. As with Logos, this revenue sustains the company and allows for more Bible engagement in the future.

YouVersion, however, has no direct transactions with customers and no ads on its website or in its apps. But the company does regularly report on its blog, through email newsletters, and in other outlets how much Bible engagement is happening, using metrics like millions of users, millions of downloads, or billions of minutes spent reading.[41] This reporting is presumably designed not only to encourage their users, but to validate the company's work to major donors, who function like customers. The push for donations by users, which is more recent, shifts this relationship and frames the transaction not as a purchase for oneself but as part of the larger mission of sharing scripture with the world. With Logos, the relationship is more analogous to print, where the paying customer is the Bible reader and personal experience with the product determines whether the customer will return or choose another product. With Bible Gateway, there is a similar connection with the reader, but the actual paying customer is the advertiser, whose desire is that Bible readers will not only see an advertisement, but click on it and purchase a product. Similarly, YouVersion has a direct relationship with its reader, but the paying customer is the donor who sustains the operation.

Stewardship and Success Are Interwoven

One might be tempted to think that the development teams are naive about the ways in which their spiritual and financial goals align or would attempt to downplay the latter. But instead, the teams were well aware of these complexities and embraced them as part of their broader mission. When I asked the Bible Gateway team how they measure success, three team members each pointed to a different measure. A programmer answered in terms of Bible engagement and enjoyment:

> Success for Bible Gateway is, I guess from my perspective, people using Bible Gateway and engaging with the scripture and enjoying it. Getting God's Word—I think that's success right there.

This response did not define "engagement," nor was it an expression of Bible Gateway's business model. Rather, it seemed to express the basic evangelical belief in the importance of connecting with God through scripture. He also emphasized that reading the Bible should be "enjoyable," perhaps echoing the contrast between "religion and relationship" commonly made by evangelicals. Another team member connected success to the company's business model:

> As a for-profit company, we do care about ad revenue and people buying products from our affiliates on Bible Gateway. So making money is probably, certainly, a measure for our success as well.

This acknowledgment of a profit-driven motive was followed quickly by another developer who connected Bible Gateway's success to a larger goal of reaching people around the globe without access to a printed Bible in their language or who were not born into the Christian faith.

> One thing that I kind of had more of a personal connection with, because it's something I had spent a lot of time working on, was, I think, our global reach. . . . It's rewarding to think about a group of people that wouldn't be

> able to find scripture in their language and on the internet before, and they can now . . . . When you think about when Jesus talked about what things are going to start looking like and about the gospel going to every corner of the earth, I feel like what we do is maybe playing a huge part of that in the twenty-first century.

This discussion moved from an overall ideal (engagement) to the financial model (advertisements) and back to the broader mission (outreach). One might presume that the third movement was a kind of justification of the second point, but the developer's tone did not suggest that he was uncomfortable talking about the need to generate income. That is, the shift from ad revenue back to evangelism was not a way of deflecting from the financial reality. Instead, this transition seemed natural, and he was excited to see the connection between engagement, outreach, and funding. When the number of page views increases, there will be revenue for the company to continue its work encouraging more engagement and deeper outreach.

The Faithlife team was also comfortable talking about building profitable tools. When I asked about whether they would consider building additional apps or features might encourage new forms of Bible engagement, one senior leader responded by saying:

> If we can make the economics work, we'll probably build just about anything. We think of that as a stewardship issue, that what we have is an opportunity to serve the church, to serve our market, and to glorify God, and we have a certain amount of resources that we have to put toward that activity.

Here again, we can observe the direct connection between creating functionality that increases engagement and building systems that increase sales. At the same time, while "economics" is a reality for a for-profit company like Faithlife, its Christian identity also means that developers simultaneously see their work in terms of service to the church and ultimately to "glorify God." They view revenue from sales not in terms of personal or corporate financial gain, but as a spiritual "resource" that they are responsible to "steward" for the purpose of encouraging more people to engage with the scriptures. Another Logos employee added that profit was a means to an end, not motivation in itself: "We could be making banking software or insurance adjusting software . . . and make a lot more money . . . than the Bible, but this is meaningful." It was their responsibility to steward what God had given them by using every tool available to encourage more engagement and more sales.

For example, Faithlife tracks how often customers use their products, and when the detects a decrease, it triggers an email to the customer offering discounts, new products, or new features, depending on what has been algorithmically determined likely to bring the user back into engaging with the software and with scripture.

> Those [email acquisition techniques] have been actually pretty successful. We do *reconvert* people now and then. . . . They buy a store product, or they get it for gifts, and six months later they still have that desire to do Bible study. (Emphasis mine).

In this context, the language of "conversion" also takes on a dual meaning, as it moves from religion to sales and back again. One of Bebbington's four pillars of evangelicalism is the concept of a spiritual "conversion," or the experience of being "born-again," which represents a spiritual transaction where nonbelievers take a step (generally through prayer) that transforms them into a genuine Christian believers.[42] This transactional sense of "conversion" was later picked up by the technology sales world and incorporated in its vocabulary for defining the success of online products and services. Companies engaged in e-commerce compete on their ability to deliver maximum "conversions" (sales transactions), and consultants offer services guaranteed to "increased conversion rates." Here the term "conversion" (or "reconvert") has moved from the religious realm into the business realm and then been reabsorbed into the sales of digital religious products, used fluidly as both a commercial and a religious activity. This appears to be another example of Connolly's resonance machine, which includes not only evangelical companies adopting business language, but the business world repurposing religious language in a secularized world.

YouVersion, as we have seen, has a very different financial model in that it has been largely sustained by donors rather than advertisers. But some of its features indicate that there is a deeper connection between its funding model and evangelical publishing businesses. On the Home tab, YouVersion displays content from what it calls "partners" such as devotional Bible reading plans developed by popular Christian authors and videos from musicians, teachers, and other ministry organizations. As I understand this model, the partners either offer their content to YouVersion for free in exchange for a link to purchase their products, or they pay YouVersion to have their content featured in the app. In this model, first a ministry receives a donation to create and distribute Christian content, and then some of that donation is funneled up to YouVersion for the distribution of the material. YouVersion reports back to the ministry how many people have engaged with the content (this number is also publicly displayed within the YouVersion app next to each reading plan), and then the ministry can report that number back to its donors. This model demonstrates another way in which the religious and social web around YouVersion and other Bible apps is very complex and interwoven with other aspects of evangelicalism. The model is not simply software creator and end user. Instead, there are multiple layers within and around the organization, influencing which features are prioritized and which ideals might never be implemented. There is also a recursion between engagement and sustainability, where YouVersion determines how successful content from its partners is by looking at whether it increases Bible engagement, and that engagement is likely to in turn produce sales for the ministry that produced it. These sales then allow the ministry to produce more devotional material, hoping, together with YouVersion, to change more lives.

The Digital Bible as Evangelical Mission

In this chapter, we have taken steps toward answering the key questions of how Bible software developers are influencing the reading behaviors of Christians by looking at how their evangelical identities, from the developers up through the management, contribute to the social shaping of three of the most widely used digital Bible platforms. We have seen that the interviewees had deep relational and experiential ties to evangelical institutions and churches that led them to adopt an understanding of the Bible that prioritizes increased scripture engagement with the goal of leading to spiritual progress and life change. They also embodied the HEP orientation to technology, business, and ministry, combining their open attitude toward forms of Bible engagement and their optimism about new technology, giving them freedom to incorporate a variety of streams into their development process. This includes their own theological instincts, as well as the behavior of their users and ideas from the wider context of the technological world around them.

We also saw that the business model for each company plays a significant role in what they do and that, though the models differ, they share a common resonance between Bible engagement and financial sustainability. But rather than functioning with the cynical instrumentalism that we see in politics, the development teams openly connect Bible engagement and business income under the rubric of biblical stewardship, seeing themselves as following Jesus's direction in the parable of the talents (Matthew 25:14–30) and Paul's example of funding his ministry of the gospel through both donations (Philippians 4:13; 2 Corinthians 2:9) and the business of "tent making" (Acts 18:3). Brandon Donaldson, pastor at Life. Church and director of their Open Digerati initiative, summarized this approach at a conference in which he said, "YouVersion is not a technology; it's a Bible engagement mission."[43] In the following two chapters, we will see the other side of the social shaping paradigm, looking more closely at the ways regular Bible readers use Bible software in their spiritual practices and how this influences their relationship with scripture.

6
A Portrait of Evangelical Bible Readers

In this chapter and the one that follows, we will begin to examine the user side of the social shaping of Bible technology, shifting our attention from the way evangelical biblicism and Hopeful Entrepreneurial Pragmatism (HEP) function among developers to the way the products they created shape evangelical readers.[1] We will see that Bible software replicates and upgrades some functionality that exists in printed Bibles, and that, if we use the language of "technological affordances" developed by sociologist Ian Hutchby, it also enables new forms of Bible engagement.[2] These changes require evangelical readers to learn how to negotiate decisions about which medium they should use for different purposes or in different situations. We will see that, on the one hand, Bible software has allowed readers to engage the Bible more often and in more forms, but the data also suggest that, in some cases, using Bible apps works against the values of the evangelicals using it. We will also be forced to wrestle with the question of the extent to which we can attribute these changes to Bible software itself (and the evangelicals who make it) or to the hardware on which it runs, particularly the phones that are now a fixture in twenty-first-century Western society.

The participants in this study displayed diverse opinions and behaviors, but they also had two important things in common. As evangelicals, they shared a deep regard for the scriptures in their lives, reading the Bible at much higher rates than most practicing Christians. And, to varying degrees, they also displayed elements of HEP in the ways they talked about technology and navigated choices about Bible media. We will see that there is a complex relationship between the hardware (a phone), the software (Bible app), and the environment (social situation), all of which affect the choices of the participants. At times, participants would be very reflective, discussing how they chose a medium (print, phone, or computer) based on how it might help them perform a desired type of Bible engagement (devotional reading, study, audio, etc.). But the same person might also mention that in other cases, they choose between print and digital formats based purely on convenience, selecting what might be called the NAB version, or Nearest Available

People of the Screen. John Dyer, Oxford University Press. © Oxford University Press 2023.
DOI: 10.1093/oso/9780197636350.003.0006

Bible. This pragmatic decision-making mirrors the words of an interviewee in a *USA Today* article on digital Bibles who said, "The best Bible is the one you have with you[, and] I always have my phone with me."[3] Others, however, had tried using their phones as digital Bibles, but ultimately found it too distracting or difficult to use. Still others preferred digital media for their personal reading but refrained from using it in worship settings or with their children.

Thus, in their decision-making, the participants revealed a complex heuristic driven by the individual's values, the properties of the media or device in question, and the environment or goals in the given situation. Layered onto their decision process are several latent evangelical values, including the significance of knowing exact Bible references, the printed Bible's ability to serve as an identity marker, and the belief that the meaning of the text can be definitively known if one reads it correctly or has the right resources.

Focus Group Approach and Summary

As outlined in chapter 2, the research on end users involved five focus groups at three evangelical churches near Dallas, Texas, and incorporated both qualitative and quantitative methods. I gave the participants an overview of the research and then asked them to engage in four activities. First, they filled out a survey that asked them about their Bible-reading habits and the forms of media (print, phone, tablet, computer) that they used for various forms of Bible engagement (devotional reading, study, etc.). Second, I split the participants into two equally sized groups—a smartphone reading group and a printed Bible group—and asked them to read the Epistle of Jude and complete an assessment on their comprehension of the passage. Third, I gave them instructions for engaging in a ten-day reading plan of the Gospel of John using the same medium they used for the comprehension assessment and reporting back afterward with the results of their reading.[4] Finally, the group members participated in a discussion where I asked them questions about the place of the Bible in their personal and corporate faith and the ways in which digital media had come to influence their individual and social behaviors around the Bible.

The first church I approached, City Bible Church, allowed me to conduct research with three "Bible Community" groups, each of which had approximately thirty-five men and women in attendance.[5] In the first and second groups, the adults ranged in age from twenty-five to forty-five, while the third group had a slightly older demographic ranging from thirty-five to sixty. The second church, Petra Community Church, had a slightly larger group of forty-five participants ranging from forty-five to seventy years of age. Finally, Hidden Baptist Church had approximately forty participants with the widest age range, from sixteen to eighty. Of these 180 people in all five groups, 150 completed all the assessments with usable data.[6] In total, the participants were 49% male, 51% female, with a mean age of 45.86 and large clusters in the 0–45 age range and 55–65 age range. When asked how many years the respondent had been a churchgoer or attended church, the average was 35.52 years.

In this chapter, we will draw from the survey data on respondents' preferred media and from group discussions in which they explain why they make these choices. In the following chapter, we will consider the quantitative results of the comprehension and daily-reading exercises, along with relevant discussion points that add color and depth to the results. What we will find is that although the participants in this study are regular attendees of evangelical churches and adhere to evangelical values such as the importance of regular daily reading of scripture, they also reflect a range of behavior patterns and opinions on print and digital Bibles. Their statements range from nuanced reflections on using technology to those that lean in a more pragmatic or experimental direction, demonstrating a willingness to try new things or use whatever is most readily available. Some respondents placed more weight on their individual goals and preferences, while others seemed more attuned to social dynamics, including those in a worship service, small-group Bible study setting, or with their children at home. Although they all came from evangelical backgrounds, they presented an array of personal and socially constructed values, and their answers reveal that multiple elements are working simultaneously in the religious social shaping of technology. Because of this, we will begin our discussion of Bible software by examining the Bible in its social context.

Social Context: Relationships and the Bible

Bible Reading Frequency

The first question on the survey asked participants how often they read the Bible and how often they would "ideally" like to read the Bible. They demonstrated their evangelical ideals, with 94% of the respondents reporting that they should read the Bible daily. However, when asked how often they "actually" engaged with the Bible, the numbers were lower. Only 41% said they did, in fact, read the Bible daily. Another 46% said they read the Bible at least a few times per week and 9% at least once a week, meaning that more than 95% of the participants read the Bible at least weekly. While this may fall short of their ideals, their reading rates are still much higher than the rates for both average Americans and practicing Christians.

Barna found that only 8% of Americans and 28% of practicing Christians reported reading the Bible daily, 8% and 28% respectively several times per week, and 6% and 13% respectively once per week.[7] Similarly, the American Bible Society's State of the Bible found that only 15% of participants reported reading the Bible daily, 13% a few times per week, and 9% once per week.[8] In Table 6.1, the data from this study are compared to data from other studies.

It is important to acknowledge that the reading rates in all of these studies are self-reported. However, the question distinguishing between ideal and actual reading rates was designed to mitigate aspiration reporting and allow the kinds of qualified responses that Malley found.[9] Even with this distinction between ideal and actual, the data indicate that the participants in this study read the Bible at much higher rates than Americans overall and more than other practicing Christians. These American evangelicals appear to correspond with the Centre for Digital Theology's findings on "Bible-centric" British millennials, who tend to engage the Bible at much higher rates than less religious readers.[10] These above-average rates of reading may be attributed to the fact that in addition to attending churches within the evangelical tradition, these participants were also regular attendees of Sunday classes. This indicates that they have a higher level of involvement than "practicing Christians" in general and that higher involvement with church may correlate with more regular Bible reading.

Table 6.1 How Often Should You Engage the Bible in Some Way?

	Ideal	Actual	ABS: State of the Bible	Bible in America: Practicing Christian	Bible in America: Average
Daily	94%	41%	15%	28%	8%
A few times a week	5%	46%	13%	28%	8%
Once a week	1%	9%	9%	13%	6%
More rarely	0%	4%	63%	31%	78%

The Bible in a Relational Matrix

The survey data indicate that the participants in this study are committed, Bible-reading evangelicals. But in the focus group, the participants revealed that their relationship to the Bible goes beyond simply reading or studying it. Many participants spoke of the way in which the physical presence of a printed Bible played an important role in their individual histories and in their social connections, functioning in part as a powerful religious identity marker. In contrast, when they spoke about using Bible software on computers, websites, and mobile devices, these types of connections were discussed far less often, indicating that there is a wide gap between receiving a printed Bible as a gift and having someone show you how to install a Bible app on your phone. In addition, carrying and displaying a physical Bible outwardly marks one as a Christian in a way that having a Bible app on a phone or tablet cannot. Similarly, handwritten notes collected over time portray a certain kind of long-standing Christian character and wisdom that simply having access to study notes cannot. This resonates with work showing the importance of the materiality of the Bible in liturgical and memorial contexts, and many of the comments of the participants echo these findings.[11]

For example, when asked why they preferred one medium over another, many participants answered in terms of features, convenience, and functionality, but one participant expressed her preference for her print Bible in terms of how it connected her to a significant relationship:

> One of the things I like about my Bible is that it was given to me by someone, and so it feels like, reading it, there's a relationship there with that person as I'm reading.

This comment is representative of what several others indicated through their own experiences. For all the importance members of the focus groups placed on the tactile nature of print, the power of Bible software, and the convenience of the mobile phone, they also weighted the significance of these differences based on how they experienced them within their matrix of relationships. When discussing which apps they preferred, their conversations often took on the familiar tones one might hear regarding any consumer product such as a car or a microwave. However, when asked about their first experiences with a printed Bible, their tone was more wistful, peppered with personal anecdotes. Some users also had personal connections to the first experience with Bible software, but when asked, "Does anybody remember the very first Bible that they received?"[12] an overwhelming number could vividly recall receiving a Bible as a child from a close relative or as an adult from a close friend or romantic partner:

> When I was in fifth grade I got the Bible at my church for perfect attendance for one year.
>
> I received a New Testament Bible when I joined the air force in 1955. I still have that Bible.
>
> [My boyfriend, now husband] gave me my first Bible when we were dating.
>
> It was given to me for free in college because I didn't have one of my own.
>
> From my mother, when I went away to college.
>
> It was from my grandmother. I just wanted to have one and she gave it to me for Christmas.

Across every group, age, and gender, these kinds of stories powerfully connected the participants to their printed Bibles. In many cases, they continued to use this first Bible into adulthood, but even those who no longer used that particular Bible for their regular reading continued to keep it as a cherished heirloom that held significance beyond the words within:

> I got a King James Version from my aunt. It's a white Bible with a zipper on it and it was for my confirmation. . . . [I was] seven or eight. [laughter] Yeah, it's the one I still use today.

Based on scholarship studying the materiality of the Bible, the deep significance respondents placed on these early Bibles was expected. What was less expected was that when I asked a similar question about their first experience with Bible software, many in the focus groups offered quizzical expressions and could not recall when they had first used an application. They could recall in vivid details the first printed Bible they had received, but it was as if Bible software was something they were vaguely aware of which had always been around in the background. However, the relatively few people who could recall their first exposure to an electronic Bible often placed the experience in the context of a significant relationship. For example, one adult participant recalled the excitement of her father, a pastor, when he used a desktop Bible application for the first time, some twenty years earlier:

> My father was a minister and I can remember in the nineties when he was getting used to a computer. He was so excited because there were these new Bible things that you hit a verse and type something in and it would print something off, so he could actually print his verses for his sermons and that was a big excitement instead of having to constantly flip from one side to the other of the Bible. And that was a big deal that I remember.

Others mentioned that their first use of Bible software happened more recently when their children or spouses showed them the software and urged them to try it:

> Our kids used Blue Letter Bible at school with their Bible study, and it's interesting. They have to click in and it takes you into like the Strong's version and roots and all that. And I would never have had that at that age.
>
> My husband's a techie guy. He's IT, so he would tell me about the newfandangled things that he would find at work and what people were using. And I think that was actually in the nineties when we first found out about [Bible software].

One man mentioned that he had previously been aware of Bible software, but it was not until he saw members of his Bible study using their apps to do things more quickly that he felt compelled to try it for himself:

> I was going to the men's Bible study with my old [printed] Bible and I'm, of course, having to hit the table of contents to tell me where the page was for the book. All of the other guys [using phones] were already reading it. I was like, okay. I've got to step into the digital age.

At first glance, this man appears to be attracted to using Bible software for its speed and the ease of navigating to a passage. And yet a closer look reveals that the impetus for trying a Bible app for the first time was not merely the merits of the device or abilities of the software, but something else altogether, namely the social pressure from the group and the desire not to fall behind, either spiritually or technologically. Downloading a Bible app or, as he put it, "stepping into the digital age," functioned not only to help him engage with the Bible, but to keep him within the norms of that men's group. This serves as an example of how the negotiation process of choosing Bible media is never one-dimensional, but tends to involve a complex combination of hardware, software, and social environment.

Forms of Bible Engagement by Media

Most of the participants in this study had been using both printed Bibles and Bible apps for many years, which meant that the novelty of using them had worn off, and they had developed ideas of which media they preferred to use in different situations. Following the question of reading frequency discussed above, the participants were asked to record which media they used for different forms of engagement and in different environments. They were presented with a table that had media along the top and forms of engagement along the side, and they were asked to check any combinations they used. They were not asked to quantify how much time they spent on using any one medium or form of engagement (other than "Long Reading"), only to indicate they used it regularly. Table 6.2 presents the results of this survey in table form, and Figure 6.1 presents a bar chart that makes some of the comparisons clearer.

The summary data in Table 6.2 show that print accounted for 49.2% of all engagement, smartphones came in second at 29.5%, and computers trailed behind with 11.2%. Although tablets and e-readers match the form factor of a traditional Bible most closely, participants reported using them the least, at only 10.0%. A Pew Research Study on device ownership in the United States released contemporaneously with this study indicated that roughly 73% of US adults own a desktop or laptop computer, 68% of Americans own a smartphone, and 54% own a tablet.[13] This indicates that the participants are choosing to use their smartphones for Bible engagement at a much higher rate than the ownership statistics would suggest. What is not clear in Table 6.2 is that tablet usage is more idiosyncratic. The few who reported using tablets tended to report using them for many different forms of engagement, while other participants used a combination of print, smartphones, and computers, but never tablets. In Hutchings's

Table 6.2 Media Usage for Forms of Engagement

	Print	Phone	Computer	Tablet
Long reading	66.0%	22.0%	6.7%	11.3%
Devotional	48.7%	45.3%	14.0%	14.7%
Study	65.3%	6.0%	18.7%	8.7%
Memorization	49.3%	17.3%	2.7%	6.7%
Meditation	53.3%	16.0%	4.7%	5.3%
Prayer	42.0%	18.7%	4.7%	7.3%
Audio		31.3%	9.3%	7.3%
Searching	38.7%	38.7%	34.0%	16.0%
Notes	78.0%	18.0%	4.0%	10.0%
Lectio divina	8.7%	4.0%	0.0%	0.7%
Totals	49.2%	29.6%	11.2%	10.0%

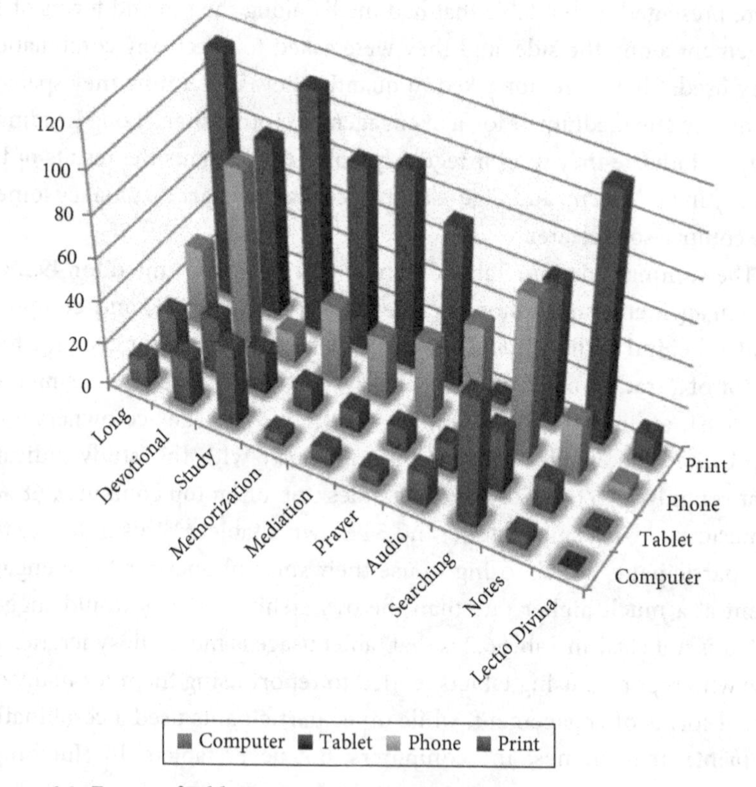

Figure 6.1 Forms of Bible Engagement by Medium

work, participants who reported using a digital Bible did so nearly exclusively,[14] but here we see only some users gravitating toward a single medium, while others took a multimedia approach. This difference may be partially explained in that Hutchings's survey was distributed online, while the present work was conducted in churches, which may have reached users who are less technologically inclined.

When it came to specific forms of engaging with the Bible, this audience reported that their most common forms of engagement were long-form reading, devotional reading, studying, and searching. Less common were prayer readings, meditative readings, and note-taking. The least common activity was *lectio divina,* which is not surprising given this audience's evangelical leanings, where this practice is less well known.[15] The data also indicate that the participants use certain media more often for particular forms of engagement. For example, computer usage is relatively low overall, but it is high for searching and studying. Smartphone usage is lower than print for long-form reading and study, but almost on par with print for devotional reading. Smartphones are also the most popular method of consuming audio Bibles.[16] Print is by far the most popular media for note-taking, with higher rates than even reading, while computers, which have the most robust ability to enter text, are the least used for note-taking.

Figure 6.1 shows that the participants continue to use print (purple) for the majority of forms of Bible engagement. However, the peaks and valleys also make it easy to see that they tend to use particular technologies for specific tasks. For example, there are phone peaks (green) on devotional and searching, while the computer peaks (blue) are on study and search. This indicates that while evangelicals tend to speak of technology in instrumental terms as being "just a neutral tool," the participants also seem to recognize the values inherent in the technology of desktop computers and phones and make media choices based on these values. We can also observe the ways in which the technology itself and the capabilities of the Bible software overlap. For example, the users report using desktops for searching, but they were not given a chance to indicate whether this was because of a desktop computer's larger screen size or the more powerful search features often present in desktop Bible software like Logos. Similarly, the users report using their phones for devotional reading, and in the discussions below, we will see that this is because of both the properties of the phone (i.e., it is always with them) and features of the application itself (e.g., daily reading reminders).

Table 6.3 Media Usage by Environment

	Print	Phone	Computer	Tablet
Home	60.0%	34.0%	22.0%	18.0%
Work/school	14.0%	42.7%	20.7%	7.3%
Sermon	48.0%	31.3%	1.3%	6.7%
Group	50.7%	36.7%	0.7%	9.3%
Kids	42.7%	12.7%	2.7%	4.0%

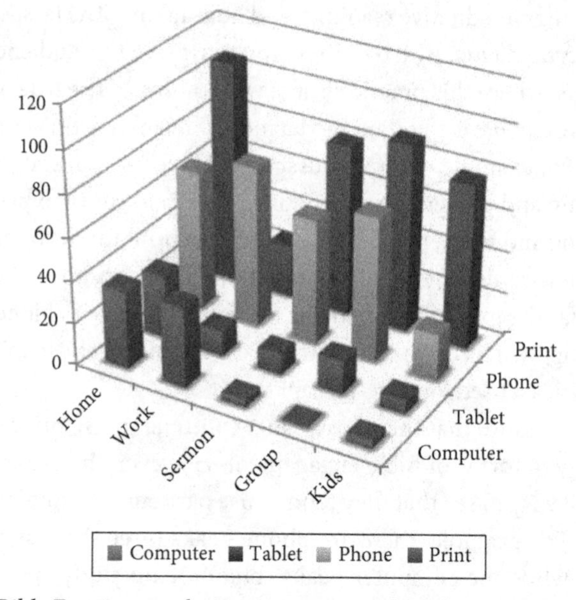

Figure 6.2 Bible Engagement by Environment and Medium

Participants were also asked about the media they preferred to use in different environments (Table 6.3, Figure 6.2). Their answers to this question presented a shift from how they negotiate Bible media in their private religious practice to how they navigate this process in social environments. Again, tablet usage was relatively low, but highest at home. Interestingly, while print was still the most common form of media at home, smartphones and computers were more common at work or school, where a printed Bible might stand out. Also, smartphone usage was stable in all environments, with several participants sharing stories of reading the Bible while waiting in a queue, but usage dropped off with children. Even ardent digital users who felt comfortable using devices in worship services and small-group Bible studies seemed to prefer print over phones, tablets, or computers when reading with children.

From these data, we can see that many of the evangelicals in this study have openly embraced technology, and that they have been willing to experiment with Bible apps to see what works for them. In the discussion below, we will add nuance to these aggregate data by exploring how they understood the features and functionality of Bible software.

Software: Replicating, Upgrading, and Enabling New Forms of Bible Engagement

The features of Bible software can be categorized into three overlapping categories: those that *replicate* engagement done in print, those that *upgrade* or extend something from the print world, and those that *enable* a new form of engagement. For example, simply displaying the text of the Bible on a screen can be understood as largely replicating the functionality of a printed book, but the search functionality of a Bible app is much more powerful than using a printed concordance and upgrades the experience. Similarly, adding a highlight color in a Bible app is very similar to what one does in print, but listening to an audio version of the Bible while one reads, goes for a walk, or commutes to work can be understood as a smartphone app enabling a form of engagement that was not available before. These categories are rather loose, and there is overlap between them, but they can help distinguish the ways the users in this study make choices and form their preferences.

Here we will focus on how the participants responded to three modes of engagement that roughly fall into the taxonomy above: search and study (upgrade), note-taking (replicate), and audio-listening (enable). In the case of searching and studying, we will see that Bible software extends and augments existing study practices, and as a result, almost all users appreciate the functionality. However, in the case of note-taking, which in many cases merely replicates the experience of print, users were more divided, with some enjoying new digital functionality and others strongly preferring print. In a third example, the audio Bibles available in most mobile apps, we will see that Bible software seems to be enabling a new form of engagement by tapping into the affordances of the mobile phone. This will lead us to consider the significance of the hardware itself, which the users primarily frame in terms of convenience.

Search and Study: Upgrading Bible Study

As we saw in chapter 3, the Bible study is one of the most important activities in the social life of evangelicals, and the data presented here indicate that many of the participants use their phones and computers for the forms of engagement labeled "study" and "search." Before Bible software, a person needed to use a printed concordance to find all the passages where a particular word appeared, but a computerized search offers a significantly upgraded experience. This includes not only the quantity of data returned, but also the experience of being able to immediately view the surrounding passage and related words. In this sense, search and study could be considered both upgrading the experience of print and enabling a new form of engagement. The significance of this can be seen in the way the participants answered questions about their first experience with Bible software. Their attraction to Bible software was not simply seeing the text of a book replicated on screen. Instead, many users recalled the first time they saw the speed and ease of using an application to perform a search that would have taken hours if done manually. One woman reported her first experience with a Bible website:

> The first thing I ever used was Bible Gateway, and I've probably used that for ten years or something, just to get a search on keywords or to look for verses, or find out how many times a certain word comes up just for different studies.

The speed and comprehensiveness of these searches appears to be a primary driver toward using a digital Bible. Several participants used a form of the word "quick" to describe what first drew them to Bible software, helping them to "accomplish" their goals:

> I was astonished at how much quicker that I could accomplish what I was trying to do because I could search.
>
> I love being able to grab my phone and go to Bible Gateway and do a keyword or passage lookup very quickly.

This corresponds to one of the major behaviors that the Bible Gateway team identified in its users, called a "quick hit." But beyond speed, participants offered several other reasons why the functionality of search was important to them. Some mentioned wanting to find out "what the Bible says about"[17] a particular word or subject. Others used the search function for locating passages of which they had a vague memory but could not remember the exact wording or location. Bielo observed that members of evangelical Bible studies often cite verses without recalling the chapter or verse, and in this study, a teenager mentioned how Bible apps have made inaccurate or false citations a less frequent occurrence:[18]

> I think it's also harder to get away with, "There's that verse but I don't know where it's found." You can just search for the keyword, and so I think it does allow in conversation to be more accurate with just talking about generally a concept from the Bible to an actual scripture reference.

The young man seemed to have adopted an idea common to evangelicals that objective, propositional truths are highly valued, and that being "accurate" is an important component of social interactions around the Bible.[19] He went on to describe how his use of a phone-based Bible application was sometimes looked upon negatively by his parents, but when they found themselves in a situation where they needed to use a printed concordance to find a passage, the parents conceded that the digital version was better because it provided faster, more accurate, and more comprehensive results. The importance of finding an exact reference in a social situation or in personal recollections was also mentioned by participants in other groups:

> One thing I do love about the digital is that there's that verse somewhere with those words, and that one about the runner or whatever, you can look it up and you'll find it, where it is and you can find it in context. Whereas if I just have my paper Bible it's like, "Ah, it's in the Gospels . . ."

This participant gives the example of recalling a small part of a passage ("that one about the runner"), but because he did not know the precise reference, a print Bible would only allow him to refer to the general location he remembered ("in the gospels"). Interestingly, however, while the word "run" does appear in several of the gospels, the most commonly referenced passages about running tend to be in Paul's epistles (1 Corinthians 9:24; Galatians 2:2, 5:7; Philippians 2:6) or the one in the letter to the Hebrews (12:1), indicating this participant might have been further away than he thought. Nevertheless, the importance of being able to find the location of the passage was again emphasized by this user. Some commentators have suggested that searching and viewing the Bible on a phone screen might lead to ignoring the verses before and after, but this participant seemed to reject these concerns, valuing Bible software for its ability to let him immediately access the "context" of the passage.[20] With a printed concordance, one only sees a small amount of text and needs to take the second, slower step of looking up the complete passage, whereas the complete passage is only a click away in a digital version. Thus, we see that, in the case of search functionality, Bible software not only upgrades the speed and accuracy available in print, it also helps users interact with the text in a more contextually rich way.

If search functionality helps users find "what the Bible says" about a particular word or topic, readers also use Bible software to help them understand the meaning of a text as they study it. In predigital Bible study settings, participants with printed study Bibles often "defer immediately to the footnote and accept its interpretation without question."[21] Historian Kathleen Boone, for example, credits the innovative use of study notes in the Scofield Reference Bible with popularizing dispensational theology.[22] In the context of Bible apps, one of the features most often mentioned was having access to commentaries and explanatory resources that provide more information than the notes in a study Bible. The following participant discussed the importance not only of having interpretive helps, but of referring to multiple commentaries whose authors wrote over a wide period of time.

> I love Blue Letter Bible. . . . I think it has the best integration with lots of options and commentary which is really nice to be able to pull from the modern commentators as well as the Luther's and old school, too.

In addition to outside resources, another theme that emerged in the discussion was that many users appreciated the way Bible software allowed them to dynamically compare multiple translations of the Bible side by side or in searches to help them better understand its meaning. Comparison was not one of the "forms of engagement" listed in the surveys, but several users mentioned using their phones to bring up multiple versions at once or comparing a version on their phone with the one in their printed Bible:

> [With my phone], I look at the different translations, going from one translation to the other.

> If I want to compare ESV to NIV to King James, it's a little hard to do that with a bunch of print Bibles out. Print is what I use for just reading. I do use my phone app with a printed Bible to compare translations.

The first speaker simply states that he uses his phone to access different translations of the Bible, and others in the focus groups echoed his appreciation for that upgraded functionality. But the second participant reflected many others who expressed more nuanced and complex practices where they valued different media for different form of engagement ("print is . . . for reading" "phone app . . . to compare"). He also mentioned using print and digital in conjunction with one another, using print as the primary source text and the phone for additional study material and translation comparison. The second speaker mentioned that he could do a comparison with printed Bibles, but that it is much easier with a phone app. Here digital media are in some sense replicating the functionality of having multiple Bible versions or purchasing a parallel edition, but this user found the app so much easier and faster to use that he considered it to be an upgrade. However, rather than using his phone exclusively, his Bible reading and study habits have been changed into a multimedia experience with his phone alongside his printed Bible.

Other participants mentioned similar patterns of using both print and digital media for comparison and study:

> I almost always work with both. . . . We were reading in Genesis and Exodus, but then sometimes they would call you to read a verse in John, and I'd be like, "It's so much work to flip some pages," but I love . . . being able to go [to] all the different translations. . . . It's something different in digital, side by side.

In this exchange, the speaker connects several themes, including speed, comparison, understanding, and moving between print and digital versions of the Bible. Again, the phone is seen to be superior with many upgraded features, but rather than shifting to read exclusively on a screen, many of the participants spoke in terms of using both side by side, taking advantage of what each offered. Toward the end of this chapter, we will continue this discussion of the participants' fluid movement between print and digital usage, but one other major study feature of Bible software is important to mention.

In the quest to understand the meaning of a passage, several participants mentioned using more complex software to study the original languages of the Bible.

> My Bible app [lets you press and hold on a word] and [it] will tell you the original meaning and stuff, and I don't care if you have big books if you have that kind of information.
>
> I love Blue Letter Bible. . . . I think it also makes really easy to go back and see what the Greek or the Hebrew says, so to really understand on a deeper level what is the actual word.

It was not clear if these individuals had prior training in biblical languages or if their flavor of evangelicalism led them to value getting to the "original meaning." In the first quotation, the participant hints that having Bible software at one's disposal means that one does not necessarily need to heed the opinions of someone with "big books." This appears to be an example of Campbell's category of "shifting authority" in networked society because the participant seemed to feel that older authorities (i.e., those who carry "big books") are no longer necessary when "you have that kind of information."[23] Instead, these apps are carrying on a tradition, set in motion with the printing press, of the democratization of knowledge. The second person also appears to believe that software could help him understand the original meaning of a text in a way that a printed English translation could not do. On the one hand, this may mean that apps are helping users understand that English Bibles are translations rather than the "literal" word of God, and yet it may also represent a twist on evangelicalism's anti-intellectualism where the app provides a veneer of knowledge and creates a sense that the right answers are (literally) at their fingertips.

These original language functions are indeed powerful, but another user highlighted the seemingly endless experience of tapping deeper and deeper to answer more questions:

> My favorite one is the Faith Library; it's loaded with stuff on there. You have the passage, then you have subtext, which I just love, things that you can buy underneath it. So you read the story and even in Jude it talks about how Michael and the Devil were fighting over Moses and it's like, well what does that mean? There's a whole book about it underneath. . . . You just click and you keep going deeper and deeper.

This participant uses several different metaphors ("subtext," "underneath," and "going deeper") to describe how the Bible software brings in outside knowledge to help the reader better understand the biblical text. As with the prior comments about clicking through to find the true meaning behind the words, these participants imply that the meaning of scripture can be known if only one has the right tools, including access to the "deeper" knowledge of the original languages.

This experience of tapping through to find answers took on another level of significance when another participant shared that he had used digital resources for a time, but then switched back to print because he preferred its "tactile" nature. However, he went on to say that as he read the printed text, he found himself "looking for commentary or clarification for certain words out of habit, and it wasn't there." His comments recall the Faithlife leaders who spoke of the poverty of print and its inability to get to help a reader get to "truth." Interestingly, this impulse (or habit) to look for meaning outside the text itself seems to run contrary to the doctrine of the perspicuity of scripture, that idea the Bible can be read plainly and understood.[24] The Westminster Confession explains perspicuity this way:

> Those things which are necessary to be known, believed, and observed, for salvation, are so clearly propounded and opened in some place of Scripture or other, that not only the learned, but the unlearned, in a due use of the ordinary means, may attain unto a sufficient understanding of them.[25]

Instead of "ordinary means," the participants above seemed to be arguing for a kind of secondary perspicuity, where the meaning of scripture can always be known, but only if one has access to the right resources with the right answers. They are confident that there is a correct answer ("What does that mean? There's a whole book about it"), but it may be in one of the "things you can buy." This is similar to Boone's argument that the Reference Scofield Bible provided a key that allowed evangelicals to "decode" its meaning. After experiencing the search capabilities of Bible software, most participants said they did not want to go back to a printed concordance, and here, it seemed that after experiencing the rich resources of the digital study environment, many users did not want to go back to only having access to the text alone. This appears to be a case in which Bible software adds a new pattern of Bible engagement to the participants' lives and partaking in that pattern also changes later encounters with the printed text and one's relationship to the

Bible itself. Likewise, while the study notes and reference material can be seen as a form of evangelical empowerment by providing access to resources that allow readers to study the text apart from an external authority, these apps could also be viewed as transferring authority to themselves, reinforcing the idea that the meaning scripture is knowable, but only if one has purchased the right resources.

However, while many of the participants appreciated the speed of digital searches and access to commentaries, others expressed concerns about these new tools. The following participant acknowledged how "technology" was clearly better in some cases, particularly for searching, but he also expressed a worry that, over time, using phones and computers might lead to the neglect of printed Bibles, and specifically the role a printed Bible plays as a social signifier of a true Christian. As the participant said the following words, he held up his well-worn printed Bible, and each time he said the word *this*, he gestured with it for emphasis:

> The technology sort of blends the Bible with all these other things. And so, when I hold *this*, people see it and they say immediately, "Oh, they must be a Bible-believing person. That person's different." And so I feel like sometimes we're losing this uniqueness that *this* represents when you hold it and when you carry it. And so technology to me is good for quick referencing and what you need in conversations when you don't have *this*, but we're losing the idea that we need to carry *this*. We can reference [the Bible] from a phone, [but] everybody else that sees that phone doesn't know. . . . So I think *this* is still unique right here.

The man praises the digital Bible for use in certain forms of engagement ("good for quick searching") and for its convenience ("when you don't have [a printed Bible]"), but also laments its uselessness relative to a printed Bible as a visible identity marker ("a Bible-believing person") and as an object with cultural and communicative power ("what this represents when you hold it and when you carry it"). Below, we will explore additional worries about Bible apps, but here it is important to note that even a person expressing the evangelistic importance of a printed Bible conceded how helpful Bible software was for searching and studying. In the next case of note-taking, we find a form of Bible engagement where there was less consensus around the benefits of digital media.

Note-Taking and Highlighting: Replicating a Practice

Unlike long-form reading, where a large majority of participants preferred print to screens, and unlike searching, where most viewed Bible software as a significant upgrade to printed concordances, in the case of note-taking and highlighting, opinions were split on which they preferred. The note-taking and highlighting features in Bible apps largely replicate the same functionality in a printed Bible, but researchers have suggested that note-viewing and note-taking features are still not as robust in Bible software as what is available on the printed page.[26] The users in this study expressed a range of opinions about print and digital note-taking, and their reasons for choosing one over the other included considerations such as ease of use, immediacy of access, and relational connectedness. For example, in terms of functionality, several users mentioned wanting their notes to be immediately visible, convenient, and accessible. The following user mentioned trying the notes feature in her digital environment, but then later realizing that she never looked at them again because the notes were not visible by default. She then contrasts this with her experience of notes or highlights added to her printed Bible, which were always visible and often prompted new thoughts:

> I prefer my own personal annotations, like I use the highlights or whatever. But then in the digital version, I don't look at them. Like I don't go back and actually open up the notes because they're not right there, [but] when I'm looking at my own notes or annotations [in a printed Bible], it's right there.

Other participants expressed similar dissatisfaction with the functionality of digital notes, but many also couched their preference for print not in terms of superior functionality, but in terms of print's perceived meaning and capacity for relational connection. For example, the following participants both drew a contrast between search functionality, which they saw as clearly more powerful in a digital environment, and the sense of meaning derived from the experience of seeing one's handwritten notes and previous highlights:

> Yeah, I certainly recognize the utility of the search function of the digital, but I want to go back into my Bible and see what I've already underlined and go back to that and all, old dates and maybe even when I shared a verse with someone else, or kind of what was going on at that time.

> I've been in technology all of my adult life. So, I'm very comfortable with it, but with my Bible, I love writing in it, notes, sticky notes, tuck-in cards, that kind of thing. I would have to say this is more meaningful to me, but I love being able to grab my phone and go to Bible Gateway and do a keyword or passage lookup very quickly.

Both seemed to view the search functionality of a phone as something that augments the deeper, more meaningful experience of print for taking notes, highlighting passages, and sharing those experiences with others. They seem to be contrasting the "utility" Bible software with the "meaning" of print, choosing to use both in tandem rather than completely converting to a screen-based Bible. Similarly, other participants framed their preference for print-based note-taking in terms of the printed Bible's place as an identity marker, which we observed in the earlier discussion about search and study functionality. The following participant considered print-based note-taking and sharing to be a "witness to others" and felt that it was such a central part of her faith that moving to digital notes would not be an option:

> My Bible is full of notes because this is my source of acquiring knowledge. God tells us to seek knowledge and acquire wisdom. So I become enlightened by someone else, perhaps, as we're studying a particular scripture and I have my note there. That refreshes my memory that I might not have remembered had I not been able to write it down. . . . This is my witness. That Bible is my witness to others. And I carry a Bible.

As she spoke, she held her printed Bible tenderly and reverently, emphasizing it whenever she referred to a note or said the word "Bible." For her, the Bible was less a conceptual category than a sacred object with meaning that went beyond the words inside. The immediacy of the notes and their permanent spatial proximity to the text made them come alive and function like a handwritten Ebenezer that reminded her of God's work in her life. A digital Bible, with its notes and highlights stored in the cloud, always accessible, but never tangible, could never leave such a lasting impression. This indicates that for many evangelicals, the Bible as a material object may continue to have enduring significance, especially if they use a phone alongside it.

And yet, while some participants above spoke of print note-taking as an important spiritual discipline they were unlikely to change, another group of participants, several of whom were younger, offered similar stories about their experiences with digital notes. The following user reported looking through all the digital notes he had made during the previous year and using it as a kind of spiritual diary:

> What I did earlier this year—I was looking back at the ones that I had highlighted, and it was really interesting seeing the date that I'd highlighted it and why I highlighted it. I went back in my head: "Oh, I was going through this." That was really interesting to see the progression of the highlighted notes, which you can't do in a book.

This use of digital notes as a spiritual record is similar to what previous participants mentioned doing with their printed Bibles. Interestingly, both groups of participants seemed to feel that what they were doing would not have been possible in the other medium. This may indicate that a person who has a highly meaningful experience with a particular form of engagement may be less likely to try that same form of engagement in a different medium, whether print or digital. This seems to correspond with Noll's idea of "culturally adaptive biblical experientialism,"[27] where evangelicals seek Bible-based experiences. They are willing to adapt to new cultural goods like phones and apps, but Bible experiences powerfully tie them to the media associated with that experience. In my interview with the YouVersion leader, he mentioned that they had considered building a feature that would create a print-on-demand Bible that would include the user's personal notes entered into YouVersion in the footer of the page, much like a published study Bible. This combination of print and digital might allow for a new form of the experiences that these participants were seeking to replicate, but so far this feature has not emerged.

In addition to the way digital notes can be used as a time-based spiritual diary, others preferred digital notes because they could access their comments in multiple places and environments. The following participant, who served as a Bible study leader at City Bible Church, was pleased to find that notes she added using a smartphone could be accessed using a desktop computer when she was preparing lessons for teaching:

> The phone has been really useful to me because throughout a week of preparation for a teaching topic, I may be listening to my devotional and I hear something, or I read something that I know will be useful for my lesson that week. . . . I'm going to highlight it [or] make a note about it. . . .

> I know I will be able to come back to it when I am actually compiling my lesson . . . whereas if I hand wrote that in the Bible I actually might not remember where they are.

Although the participants were split on their preference for print or digital for note-taking and highlighting, both groups expressed the same basic criteria for their decision—how easy it was to go back and find the notes or highlights they had made. Some moved to digital notes for fear of losing a print Bible, while others avoided digital for fear of losing all the electronically entered notes in a version upgrade or software change. It is also important to observe that in the quote above, the Bible study leader mentions "listening to my devotional" while taking notes on a phone. This highlights two significant trends. First, as we will discuss below, many participants spoke about how phones had enabled them to transition not merely from text on a page to text on a screen, but from reading the Bible to listening to it. Second, this activity of listening while taking notes on a device indicates a multimedia and multisensory experience that goes beyond the print versus screen dichotomy to a new forms of interaction with the Bible.

While I have cast note-taking and highlighting primarily in terms of replicating the functionality of print, there is another aspect of note-taking that upgrades or even enables a new form of engagement. In the previous chapter, we noted that from its origin, YouVersion's leaders felt strongly about the importance of adding social features to their website and apps because they instinctively felt it would boost Bible engagement. They have done this with their note-taking and highlighting functionality, allowing users to quickly share verses and imagery with friends in the app or on social media platforms. Several participants discussed how these social features had influenced their use of highlighting and note-taking features, including the following user, who was initially skeptical of YouVersion's social features, but after trying them found that highlighting within a social app was a positive experience:

> I thought it was cheesy at first, the whole friends thing [in] YouVersion. I didn't think that was really going to catch on, for one. But I started doing highlights and I would share them to different people. And the one person that picked up on it was my mom, who I don't talk to a lot about the Bible stuff. We actually have conversations through the app. She's not tech savvy at all [laughter]. But we actually have conversations about Bible verses now through the app.

This participant's comments seem to indicate that the social significance of a printed Bible has, for some users, found a new form in the digital world. Digital religion researcher Peter Phillips argues that in digital culture, there is a "social performance of bible engagement" that appears whenever "people highlight, bookmark or share a bible verse within social media." He argues that these actions reveal "something about their own identity and also something about their own perceived audience."[28] In this case, however, the audience (his mother) seems to have been largely unintentional. The social performance on public platforms like Twitter and Instagram may differ, then, from what occurs within closed, religious communities like YouVersion. In this case, we can see HEP influencing the social dimension of religion, where the participant embraced a feature he doubted ("I didn't think that was really going to catch on"), experimenting with it ("share them with different people") to see if it would work ("we actually have conversations").

Again, participants seemed to see the upgraded functionality of digital search in largely positive terms, but they were more divided on note-taking and highlighting. Because note-taking largely replicates an action one can do in print, the participants seemed to tie their evaluations of the media not merely to functionality, but to their lived Bible experiences. For some, these experiences happened at the individual level, such as those who discovered of a spiritual diary in their notes. For others, their note-taking and sharing was tied to the social dimension of their faith. In the next section, we will see a feature that cannot function in print media, enabling an entirely new form of engagement.

Audio Bibles: Enabling a New Form of Engagement

Although many of the questions I asked the participants emphasized textual consumption (i.e., "reading") of scripture on the printed page or on a screen, a subtheme of increased audio-listening also emerged organically in the focus groups. In the forms of engagement survey, participants were asked if they used phones, tablets, or computers to consume audio. A few people wrote in the margins of the survey that they had previously used other media like compact discs (CDs) to listen to the Bible, but many more used their phones to listen. Audio Bibles have long been available to purchase, but the convenience of the smartphone paired with free audio versions appears to be enabling a new form of engagement that was previously too costly or inconvenient for most participants. I had not planned to ask about audio Bibles, but several users brought them up unprompted when I asked about the most significant way that digital media have changed their religious habits. One older participant immediately answered, "I think the audio apps," and others offered similar praise for listening:

> I love the audio versions. I listen to them all the time [using my phone]. I'd say probably 20 or 30% of my Bible reading this year has been audio.

Their discussion of audio Bibles surfaced several interesting trends in Bible engagement, including additional support for the idea that Bible experiences are becoming multimedia encounters with a mixture of print text, digital augmentation, and audio overlay. Several users also mentioned creating what we might call an ambient Bible experience where they listen to the Bible in the background while they go about some other task, including household chores, commuting to work, or going to sleep. Some participants also reflected on listening as an important religious act:

> At night, and I'll just have the audio on. And I just want to get back to sleep and it calms me.... It's kind of interesting, the passage says, "Faith comes by hearing," not by reading the Word of God.

The passage the participant mentioned is Paul's discussion in Romans of the importance of the proclamation and hearing of the gospel message (Romans 10:17). This participant went on to mention the ministry that took Paul's words as its name (Faith Comes by Hearing) and began producing audio Bibles, first on audio cassettes, later on MP3 players, and today in Bible apps. Others participants mentioned that before the era of phone apps, they had wanted to listen to audio Bibles, but they had not done so because of the barriers to entry including cost and lack of portability. This changed with the advent of the smartphone, which opened up the possibility of listening to scripture in new situations:

> I'll hit audio on my drive to work because at that point I'm not going to be looking at it anyway. I'm taking in scripture but through a different media on my drive to work.

In this example, Bible software is not so much extending or modifying a form of engagement that the participants were already doing, but instead, it is enabling a new form of engagement that they had not been able to experience before. However, the availability of this new form can be attributed as much to the software as to the portability and presence of the phone. The phone is at once enabling the new form of Bible engagement and allowing that form to fill a space where the Bible was formerly not available.

Others mentioned having audio Bibles on in the background as an alternative to music or news radio, and another group listened to Bible recordings as a way of going to sleep. The following participant, who said that the availability of audio was the most important change Bible apps had brought into her life, described how she used it to create an ambient Bible experience where she was always absorbed in scripture even if not consciously:

> I can go about my daily routine and I'm hearing the Word, and I may be cognitively focused on something else, but the Word is always back there in my mind.

Several other individuals confirmed that they, too, enjoyed this kind of ambient Bible experience for the way it shaped their attitude in ways they considered more positive than news or other content. Others, however, reflected that this practice did not result in them remembering as much as they had hoped, but they persisted because it was still better than not listening at all:

> I listen to an audio Bible every day when I'm either working out or walking my dogs. I'm not sure if I'm retaining as much as if I was sitting down and using quiet time, but a busy life, you know?

In this user's words, we can begin to observe a pattern latent in many of the focus groups of a pragmatic approach to Bible media choices in which any Bible engagement is better than none. This participant values a separate, distinct time of reading scripture (a "quiet time" in evangelical parlance) presumably in print, but he laments that the pace of his life is too frenetic and hurried for this. His solution was to multitask, to layer audio Bible listening on top of an existing activity, in his case, dog walking. Listening to the Bible in the background in the manner the participants describe may represent a form of what linguistic scholars called phatic communication, which is "the background noise to our lives, the social radio playing in the corner, the sense that others are present with us."[29] However, it is unclear what the effect of phatic Bible engagement might be or how effective it is at helping listeners do what they say they value, in this case, to "retain." Several studies support his worry, including those indicating that multitasking may either be ineffective or that it might actively lower IQ and learning.[30] However, some studies also suggest that multitasking does not harm productivity when the two activities do not physically conflict with one another.[31] Further study would be needed to determine the full impact of ambient Bible experiences, but here we can observe that mobile Bible apps are enabling a new form of engagement and that many users have incorporated this new practice into their lives.

As with other forms of engagement, the social dimension also surfaced in discussions of audio functions. Some participants spoke of listening to scripture in groups as they read along in their printed versions, and that they valued the practice because listening "changes the way you hear it" from the experience of reading the text to oneself. But perhaps the most significant social interactions centered around audio Bibles came from parents discussing interactive audio Bible apps designed for children. In late 2013, YouVersion launched the Bible App for Kids, which contains abridged Bible stories where the text is read aloud to the users.[32] Parents in several focus groups praised the app for allowing young children who had not yet learned to read to have the experience of "reading" the Bible. Some even wondered aloud if these younger children had listened to more of the Bible than their older children whose only access to the Bible was when the parents read to them, rather than having an app that could read anytime.

> I love the YouVersion Kids Bible, and I'm thinking as I'm sitting here that my four-year-old probably reads and knows or really listens to more scripture than my twelve- and thirteen-year-olds did at that age because . . . they were dependent on me reading it to them. To me, that's a beautiful way of utilizing the technology.
>
> We have the Bible app for kids, and that's been really helpful because, despite our best efforts to read to them daily from the Bible, it's really hard—especially for kids that don't read. But the Bible app for kids—we hand them our old iPhone 3s . . . and they love it. They just have it read to them, and they think they've got this special treat because they've got the Bible app.

> So that's been really helpful because then we know that they're their daily reading in too.

Both of these parents emphasized the importance of daily Bible reading ("our best efforts to read to them daily") in their family and that "more scripture" is always better no matter the medium or the person/app doing the reading. However, recall that in the survey data, parents who used their phones at high rates for their personal reading reported that when they read to their children, they preferred to use a printed Bible. This is perhaps because they felt the printed Bible was a better means of passing on their spiritual heritage. And yet, as much as they value the parent-child connection built through reading a printed Bible together, they value the amount of reading even more, and HEP frees them to let an app read to their children. These behaviors also remind us that children today are growing up with personalized, multimedia Bible experiences that include text, images, audio, and even interactive elements. Thus, these parents are passing their values (95% said daily Bible reading is ideal), but Bible software is creating a environment different from the one the parents would have experienced in their early faith formation.

Hardware: The Convenience of Mobile Devices

In the discussions above, we have been examining how Bible software replicates, upgrades, and enables new forms of Bible engagement and how evangelicals understand and appropriate these into their lives. We have seen that the participants chose between different forms of media based on their understanding of how each medium will help them accomplish their goals for engaging the Bible. In many cases, users compared and contrasted the strengths and weaknesses of printed Bibles versus Bible software applications. But while the software clearly made a difference in many of these cases, such as the much more comprehensive and interactive search results of an app compared to a concordance, many users also seemed to indicate that they used Bible apps less for any specific feature of the software than for the convenience of always having a phone in their pocket. This meant that sometimes the participants reflect on the media available to them, evaluating and using what they consider to be the best tool for the job. But in many other situations, they are less reflective, using not what they consider to be "best" but taking a more serendipitous approach and using what is most readily available.

The following participants serve as examples of the more thoughtful approach, noting how they chose between print and digital or between different Bible apps based on the specific activities in which they are interested.

> I tend to use my phone more for daily devotionals than my printed Bible, but I use my print version for everything else.
>
> I use the computer. . . . If I want to do commentary or searching more in depth. But the phone for the actual bible study.

In both cases, the users are reporting that they have consciously thought through which medium they found most helpful for the forms of engagement they found most meaningful. Others mentioned that their media decisions were driven not only by the activity they wanted to perform, but by the location in which they wanted to interact with the Bible. The following participant linked the importance of hardware and software together, praising the app for keeping her place in a reading plan and the phone for being accessible when she was away from home:

> I use my [printed] Bible for home study, but if you're out I always use the tablet or the phone. You have to remember where you [last read when you] walk through the Bible for a year.

Other participants offered their own taxonomies of usage based on various situations and needs they had encountered. As they talked through these scenarios, a theme emerged that even though they actually preferred print *as a medium*, they still found themselves using digital media purely out of convenience and availability. One man put it this way:

> I use [my phone] more for convenience, but I love the printed page. But I don't spend time with it. It's my iPhone I use.

Here the man admits that even though he prefers print, he no longer uses it because it feels to him far less convenient than his iPhone. The pragmatic power of convenience in determining his choice can be seen in the emotional contrast of speaking of print as something worthy of "love" and yet reluctantly acknowledging that he doesn't "spend time with it." Similarly, another man confessed that it was not until the present study that he had actively considered which device he would use in which situation.

> I had to process and think about just when and where I'd really chosen to use one thing over another, [I just tend to use] whichever [is available].

This suggests that while some users choose a medium based on its functionality, for many others, their favorite Bible translation might be called the Nearest Available Bible (NAB). That is, their primary decision-making criterion is less theological or ideological than a function of what is closest and most accessible in a given situation. These comments indicating a preference for the Nearest Available Bible correspond with the survey data on the devices participants tend to use in certain contexts. For example, participants reported that their highest usage of desktop computers was during work hours. They likely used their desktop computer because it is less conspicuous than a printed Bible and was less likely to draw attention from coworkers. But the choice may also have been driven by their desktop computer functioning as the Nearest Available Bible. Similarly, users reported the highest rates of smartphone Bible usage when they were in places other than home or work, indicating again that they preferred print but used their phone because of its affordance.

The importance of portability was also raised by participants who reported that they were initially uninterested in the concept of a digital Bibles because their first encounter was on a desktop computer. When focus groups were asked to recall the first time they encountered an electronic Bible application, one man recalled being shown the BibleGateway.com website on a desktop computer and, while he found it intriguing, he was less interested because it was not portable. Later, when he purchased a smartphone, the prospect of reading the Bible on a screen became more desirable. These comments suggest that when people are choosing between print and digital Bibles, they do so in part based on the medium's inherent qualities, and one of the most important of these qualities is the smartphone's convenience and proximity, or its accessibility in a given context. In fact, of all the questions asked in the focus groups, the participants offered the most consistent responses to the question of why they preferred Bible apps to printed apps, with many people using the words "convenience," "ease," or "accessible" in their responses.

Some went on to describe the way this convenience allowed them to fill their "downtime" with Bible reading and even increase the amount they were able to read on a daily or weekly basis. Rather than simply transferring the existing habits and disciplines from print to screen, many indicated that it was the combination of hardware and software that helped them increase their Bible engagement, sometimes rather significantly:

> I have the opportunity to read more often or do my devotional while waiting in line somewhere. I don't have to wait to get home to use my Bible.
>
> For me, [Bible reading] was sporadic until I went to an electronic copy, which is where all the rest of my work is being done. Then it became daily.
>
> Because it's on my phone I refer to it a lot more often than I did before.

We have seen that for the evangelical readers studied here, "daily" reading is an important part of their religious life, so if a phone-based Bible led to accomplishing that goal, then it will certainly be seen as a moral good. Even more important was another user's statement that the Bible app was not merely a means to an end, but helped establish a pattern or "routine" in life, integrating technology and faith together. The following participant drew on elements of HEP to make a case that the Nearest Available Bible is the best Bible:

> I grew up . . . always carrying my paper Bible to church, and I have a couple that I really like. But my whole thing for me is that the Bible that you read is better than the Bible you don't regardless of what form it comes in. I'm just not carrying my paper Bible with me everywhere, every single day, at all times. I always have my phone—always—with almost no exceptions.

The ever-presence of the smartphone ("I always have my phone") combined with the pragmatic logic ("The Bible that you read is better than the Bible you don't") make the smartphone an attractive option, and perhaps even an inevitable one. Another implication of his argument is that not reading a Bible is far worse than reading in an inferior format, which is a natural outworking of evangelicalism's high regard for the Bible paired with a hopeful and pragmatic view of technology. However, this participant also went on to say that the convenience of a smartphone Bible is a double-edged sword, because the other applications on a phone have the potential to distract and compete for attention, both when one is using a digital Bible app as well as during the downtimes that many people fill with smartphone usage:

> When you have that five seconds of downtime you're pulling out your phone to get distracted by something and usually it's Facebook or Twitter. And so to try to more often open up the Bible app and make some more progress in whatever the day's reading was, that's been a huge help.

This reference to using a smartphone during "downtime" follows reports on smartphone habits indicating that many people use their smartphones in times of boredom such as in a queue at a grocery store, during commercial breaks on television, in the restroom, or on public transportation.[33] Siker argues that when Bible apps exist alongside these other activities, it contributes to the "the de-sacralization of the Bible on digital devices." But he also argues that this presents opportunities for "a new vernacular of Christian faith for reading, hearing and watching the Bible."[34] Many of the participants in this study reported attempting to navigate this new reality by avoiding "Facebook or Twitter" in such idle moments and, instead, using their downtime for Bible reading. The focus groups were also asked if they treated Bible apps on their phone differently than other apps, such as hiding them or prioritizing them, as this participant mentioned. One person mentioned replacing a specific app (Facebook) with a Bible app:

> Yeah, I deleted Facebook and I put the Bible out right where Facebook used to be. And so my friends are, "Oh, I better read the Bible [laughter]."

Interestingly, this comment surfaces the purposes users assign to their devices. The smartphone has its origins in landline phones, and, at first, their only function was to enable social interaction through human speech. But as cell phones became smartphones, they added more features and become more of a mobile computer terminal. Today, smartphones are used more often for texting, video-watching, and internet browsing than phone calls.[35] The user who deleted Facebook seems to want to demote social interaction via Facebook in favor of a Bible app, but then does so within a larger social matrix where friends see one another's phones and their choices then influence the rest of the group. The presence and arrangement of one's apps, including Bible apps, might then be functioning as a kind of conspicuous consumption that combines software, hardware, and social expectations. It may also represent part of Siker's "new vernacular," where evangelical readers appear to view and treat their Bible apps differently than other apps. At the same time, deleting an unused Bible app is not likely to have the same social impact as destroying an unused printed Bible.

Some participants also reminded the group that the importance of convenience and availability for regular Bible engagement was not exclusive to the smartphone. One man mentioned that during his time in the US military, he found himself reading scripture more often than in his civilian life. He said that his deployment was one of his "lowest points," so I asked him if these low points were what drove him to read scripture more. He agreed to that, but also cited the fact that he carried a pocket Bible with him and that its presence and availability during downtimes led him to read it more:

> Yeah, what kind of kept it on me was [a little Bible] in my pockets through the mission—things like that. So I would keep it on me all the time. And obviously, when I had some downtime, I'd go out and read it.

Here, again, the concept of "downtime," when paired with availability, was an important factor in Bible reading regardless of media. Though a military deployment is quite different from a long queue at Starbucks, the way he spoke of having a printed Bible with him "all the time" echoed how many participants spoke of their phones ("I always have it, no exceptions"). Users also mentioned that convenience and accessibility were important not only for daily or devotional reading, but also for study, search, and other forms of engagement as they become more comfortable with them. One participant added an important nuance to the discussion of attention and convenience, noting that she read the Bible more when some form of it was visible or in her line of sight regularly. She mentioned experimenting with placing a printed Bible where she could see it and then transferring this idea to her phone by putting a Bible app in a prominent place on the home screen:

> I would say a lot of times it's out of sight, out of mind. So the print version of our Bible, we try to keep it out on the table sometimes, but sometimes it makes its way to a bookshelf. . . . And then I have a Bible app to try to counteract the out of sight, out of mind thing. I try to move it to my front page. . . . If I see the Bible it's like, "Maybe I should pick it up and read it."

This participant's comments remind us that the choices users are making between print and digital media involve both hardware and software and that they take place within a complex social environment with many sources vying for attention.

The Fluid Usage of Print and Digital

In this chapter, we created a portrait of evangelical Bible readers, drawing on quantitative and qualitative data in order to see how Bible software is influencing the way they engage the Bible. We have seen that the users in this study read the Bible quite frequently, are open to experimenting with technology, and use an assortment of forms of engagement as part of their spiritual practices. They are comfortable using a several different media to accomplish their goals, and the negotiation process they use in choosing a medium is driven by three major factors: the software, which augments existing practices and provides new functionality, the hardware, which makes it conveniently available in new situations, and the social environment, which shapes the norms of a given situation.

When given the choice, many participants said that they prefer to use a combination of print and digital forms, sometimes simultaneously, and that they move fluidly between them. The following individuals said that their use of media has evolved over time as they have learned to adapt to what each medium offers:

> This may be generational completely, but . . . I use both. I use a computer every week to learn the Bible. It seems to me the print version seems more permanent. It seems more significant. Maybe that's just because that's what I was raised with. . . . If I read a passage on the phone or on my computer—I'll go to the print version just to check.
>
> I use them at the same time. And I think that one on its own generally leaves you feeling just a little bit handicapped, compared to what you can do with both together.

These verbal comments match the data in Table 6.2 as well as written comments where participants reported using multiple forms of media to access the Bible. This fluid movement between print and digital media also follows research into the online and offline religious habits of people of various faiths.[36] Rather than seeing a strict delineation between online and offline, users move fluidly between both environments throughout their daily lives. Similarly, in her book-length treatment on digital reading, linguist Naomi Baron suggests that while screen reading tends to be linked with distractions, the future of print and screen reading may come down to a complex series of user preferences for specific forms of reading.[37] Her conclusion appears to be in line with the argument here that participants consider the software, hardware, and social context when making decisions about whether to use print or digital Bibles. Some, like the participant who verifies a digital passage by checking print, leaned in the direction of print because of the feeling of spiritual connection that that medium gave them. They referred to the printed Bible as "more permanent" and "more significant" while also recognizing that assessment might in part be due to familiarity and upbringing. This displays a kind of self-awareness that one's media choices will never be entirely objective, but deeply woven into one's history and the cultures in which one has been embedded. And yet, while they prefer print, they also include digital media as part of their Bible engagement even if "just to check." Others, however, expressed that they had read the Bible more often since the advent of mobile Bible apps, and that after experiencing their powerful features, they felt as though print alone would leave them "handicapped," which again points to a reconfiguration of the Protestant doctrine of the perspicuity of scripture. The following comment expressed the best-case scenario of many participants:

> I prefer to have a trusted printed version that I can read . . . and then some sort of digital reference that I can use as a complement to that.

The Next Generation of Multimedia Bible Readers

Most of the participants in this focus groups were adults, but I also asked a group of five teenagers if they ever thought about the world before smartphones. They answered with a simple no. And yet, when I asked if they exclusively used their phone-based Bibles, one teen shrugged and offered, "Not only." The attitude of this teenager seemed to indicate that choosing between print and digital Bibles was simply something Christians do, and this everyday decision was not particularly noteworthy. This reflects research into the way teenagers make media choices,[38] but it also reflects the position of some adult participants who admitted that they had not given their choices much thought until the present study. Print, screen, audio—these are tools we all have access to, and we just need to figure out a way to use that works for us. But these tools might have a greater influence on their users than they think, and in the following chapter, we will consider data indicating that digital media might be affecting the forms of Bible engagement they value most.

7
The Influence of Digital on Evangelical Reader Behavior

The previous chapter painted an initial portrait of how evangelicals make decisions about whether to use print or digital Bibles, noting that they have a complex decision-making heuristic that takes into account the Bible software, the phone hardware, and their social environment. We saw that Bible software augmented existing practices and afforded new ways of interacting with the Bible, but we also noted that the convenience of portable devices drove evangelicals' decision to use their phone even when they felt that a printed Bible might offer a better or more meaningful experience. These conclusions were largely based on self-reported survey data and focus group conversations, and in this chapter, we will incorporate instruments designed to measure how closely aligned their impressions are with their actual activity. We will investigate two specific forms of Bible engagement that evangelicals deeply value—reading for comprehension and maintaining a daily reading habit—drawing on qualitative and quantitative data to understand how Bible software, and more specifically the evangelicals making it, are influencing the way readers encounter scripture.

We will find that, in the case of reading comprehension, screens may indeed be impeding their ability to understand their sacred text. Though comprehension is certainly not the only thing evangelical readers value when engaging scripture, it can serve as a vector for something we have previously seen is a key component of evangelical culture—the Bible study. If comprehension serves as a case where the digital Bible may be working against evangelical values, the data also indicate that Bible software increases daily reading patterns and completion rates for reading plans, which are also important in evangelical culture. The data also suggest that these effects, both positive and negative, have an unexpected gender-based difference. Before analyzing the data, we will first briefly review the data on screen-based reading.

People of the Screen. John Dyer, Oxford University Press. © Oxford University Press 2023.
DOI: 10.1093/oso/9780197636350.003.0007

The State of Reading on Screen

In her book *Proust and the Squid*, literacy scholar Maryanne Wolf argues that the human brain was not "born to read," but that it is a learned behavior and one that is changing as we incorporate screen reading into our habits.[1] In her follow-up more than a decade later, she found that even though she a reading expert, her ability to perform "deep reading" has been diminished by her use of screens.[2] Before such books, early studies comparing print and screen reading highlighted the significant physical differences between paper and the large CRT monitors of the era, but as screen technology evolved to eliminate many of the physical disparities, it has become clear that a simple dichotomy between "print" and "screen" does not fully account for how the two media influence behavior and usage.

In his definitive 1992 survey of research on large CRT screens, Dillon showed that studies leading up to that point tended to focus on issues such as fatigue while reading, scrolling accuracy, slower reading on screens, and resultant comprehension difficulties.[3] In the transition from CRT monitors to flat-screen LCDs, users experienced less discomfort with the screen itself, and some researchers found no difference in comprehension between print and screen.[4] However, other researchers continued to find that, for both consuming and producing content, users performed better with print than with screens.[5] They concluded that this difference was likely due to higher levels of stress and tiredness with screens, which lead to decreased comprehension. These initial concerns seem to have been somewhat mitigated by modern devices, with some participants reporting that higher-resolution screens made it easier for them to read than print.

> I'm just starting to have to wear readers [glasses], and I find that I can make the font much bigger on my phone or on my computer. And so if I don't have my readers nearby, I love the technology because I haven't invested in a big-print Bible because they're too heavy.

Print versus Screen: Form and Features

As screen technology continued to improve, scholars began to suggest that paper and screen will never be entirely equivalent and that it would be more helpful to focus on the unique features of each medium and the opportunities these present.[6] Four features of this research are important to point out. First, several researchers have found that the act of scrolling on a desktop LCD screen impedes both reading and writing because users have an insufficient representation of the text compared to print.[7] In other words, the physical features of a printed book make it mentally easier to navigate and comprehend than the virtual equivalents on a screen. Second, other researchers have found that specific features of screen-based interfaces can cause problems. One team found that documents with hypertext links (such as the majority of web pages) are more mentally and cognitively demanding of readers.[8] The time given to visually processing a link and deciding whether to click it impaired performance in reading and comprehension. As we saw in the previous chapter, Bible apps add many interactive elements to the text, which may lead to similar impairment of comprehension. Third, researchers have found that screen users tend to spend much of their cognitive energy and time scanning screens for user interface components, and this scanning becomes so habitual that it influences what readers do when their eyes finally meet the desired text.[9] Rather than being focused on connecting with the message of the text, those reading the Bible on screen may also find themselves similarly distracted.

The ability of screens and digital media to show more than just text, but illustrations, video, and other elements (what scholars call multimodality) appears to reduce the amount of sustained reading a person is capable of in one sitting.[10] Some participants in this study echoed this when they expressed frustrations with digital Bible interfaces, such as the following users who commented on the smaller screen size of a phone and the frustration of scrolling when reading:

> If I'm studying [on] paper . . . it's easier for you to go back and forth because there's more of it visibly there, whereas my Bible app cuts off after a chapter and I've got to swipe to the next chapter or swipe back.
>
> I won't use [a Bible app on a smartphone]. I will only use it on the computer. I don't use any of the apps, because I need a bigger screen.

By 2010, when Apple released its first iPad, research began to shift from comparing print reading with desktop monitor reading toward comparing print reading with tablets and e-readers. Again, many researchers found that print readers still had higher comprehension than screen readers, but the results were more nuanced, with several researchers finding that print and screen readers performed equally well under certain conditions.[11] For example, a Beijing research team found that readers who were more familiar with a tablet performed significantly better than those less familiar with the device.[12] Another key study found that print readers outperformed screen readers when the time was open ended, but when a time limit was enforced, print and screen readers achieved comparable levels of comprehension. The researchers concluded that people reading on a device, when given no other prompting, tended to default to a scanning form of reading that impaired their comprehension. But if factors like a time limit were introduced, it might be able to coax them into giving their full attention to a screen.[13] Does a church or small-group context provide a similar clue to Christians that helps them avoid scanning and encourages them to devote their full mental faculties to reading the Bible? A research team led by literacy professor Anne Mangen found another subtle difference between print and screen reading. Their research found that overall comprehension between print and screens was similar, but that screen readers performed worse on the specific task of ordering chronological events from a narrative.[14] This prompted me to add a question that asked the participants to order the events from what they read to see if these results would be replicated.

Reading by Demographic

Reading researchers have also suggested that age, gender, and other demographics may have an effect on reading across different media. One study of middle-aged adults found that gender and education affected print comprehension but did not affect screen reading comprehension.[15] Another study of college students found gender-based differences in media preferences and online browsing behaviors, and studies of high school students show that gender plays a role in reading frequency (girls were more likely reading for pleasure) and media preferences (girls were more likely to prefer print, while boys were more likely to prefer screen reading).[16] These studies did not suggest a difference in comprehension based on gender or media, but they do indicate that we should be careful to take these factors into account when studying Bible reading.

More recent studies suggest that the gap in comprehension is narrowing, with some researchers finding that e-ink displays are less tiresome than LCD screens, leading to better recall and little or no difference between the comprehension of print and screen reading groups.[17] These studies lead us now to consider the way screens and Bible software may affect the act of reading the Bible. The following data look at two specific activities, reading for comprehension and maintaining a daily reading pattern. We will see that the screen may be negatively affecting comprehension while at the same time helping users to read the Bible more often.

Comprehension Issues, but Only for Some

After completing the survey on reading behavior and media preferences, we moved on to the Bible comprehension assessment. Participants were split into two groups, a print reading group and a phone reading group, and asked to read the Epistle of Jude and then answer six questions designed to measure their comprehension of what they read and how it affected them. Participants who correctly answered all six questions received a 100% score, those who answered five of six received an 83.3% score, and so on. In the data below, the scores were grouped by technology, demographics, and other factors, and then averaged. In Table 7.1 below, we can see that print Bible readers scored higher (66.2%) than phone readers (59.2%), which confirms priors studies concluding that print users generally have higher comprehension than digital readers. However, it is important to note that the difference between these results is not statistically significant according to the conventional measure ($p = .19$), and that larger study groups of participants would be needed to verify the gap in comprehension observed here.[18] It would also be interesting to repeat this kind of test outside of a church environment to see if the religious setting played any role in the difference.

I also analyzed the data by age and gender, where more statistically significant differences emerged (Table 7.2). Participants over thirty-five and those under thirty-five both scored higher when reading in print than on phones, and those over thirty-five scored higher than those under thirty-five. Biblical

Table 7.1 Print versus Smartphone Bible Comprehension Assessment by Gender

	Print	Smartphone	p-Value
Overall	66.2%	59.2%	.19

Table 7.2 Print versus Smartphone Bible Comprehension Assessment by Gender

	Print	Smartphone
Over 35	67.0%	60.2%
Under 35	62.5%	55.6%

Table 7.3 Male versus Female Comprehension Assessment by Medium

	Female	Male	*p*-Value
Overall	62.6%	60.2%	.61

Table 7.4 Print versus Smartphone Bible Comprehension Assessment by Gender

	Print	Smartphone	*p*-Value
Female	62.5%	64.2%	.56
Male	69.8%	53.3%	.07

literacy data from the American Bible Society suggest that although younger generations are less biblically literate overall, among those who are committed Christians, biblical literacy for millennials is only slightly lower than Gen-X or Boomers and in some cases higher than Elders.[19]

While the aged-based comparisons were not all statistically significant, the gender-based comparisons were. Female participants scored slightly higher than male participants overall when combining both print and screen scores (male: 60.2%, female: 64.2%) (Table 7.3), but things get more interesting when the results are split into print and screen. Here we see that female participants scored almost the same reading print and smartphones (print: 62.5%, digital: 64.2%), but there was a much wider and more significant gap in scores for male participants (print: 69.8%, digital: 53%, $p = .07$) (Table 7.4). This indicates that the differences in comprehension by medium were primarily driven by male participants. Discussion of this gender-based difference will follow in the sections below after we consider a similar difference in the daily reading assessment.

From Comprehension to Hermeneutics and Experience

The assessment had six questions, the first four of which were used to calculate this score. The questions were designed to assess whether the readers could recall information presented in the text, and the answer choices were intended to prevent the need for hermeneutic or interpretive ambiguity. For example, two of the questions asked the following:

> Jude mentions that he has a brother. What is his name?
> ❑ James ❑ Peter ❑ Andrew ❑ Thomas

> Michael the archangel and the Devil contended over whose body?
> ❑ Enoch ❑ Elijah ❑ Moses ❑ Christ

A third question asked participants to recall a portion of the text of Jude correctly. The fourth question was not multiple choice, but instead asked the readers to correctly identify the order of three of the five Old Testament stories employed in Jude's argument. Similar to research indicating digital readers had more trouble ordering events in a narrative, the phone readers in this study placed the events in the correct order 32.9% of the time, while the print users were correct more often, at 41.4% of the time. Knowing the order of events is not necessary for religious experience, but often is an important element in biblical narratives, and it appears that even for short texts like Jude, scanning behaviors on screens might inhibit retaining this level of the story.

In contrast to these objective questions, the fifth question asked the users to write a short response to this prompt: "What point is Jude making by bringing up these Old Testament stories?" Their open-ended answers could be grouped into several distinct categories. Some only offered short statements like "Do better," "God's wrath," and "I don't know," while others offered longer, more nuanced statements. The majority of longer responses emphasized God's judgment and the presence of evildoers with comments such as "Evil men and those who turn away from following God are always at work + trying to lead others astray" and "Remain faithful in God's teaching or you will be judged by God and punished." A smaller number of participants focused more on God's faithfulness or kindness to his people with comments such as, "To show how God provided through the generations for his people." Some also combined these themes: "God rescued and redeemed the faithful or his people but he also consistently punished/punishes disobedience. He is righteous." Interestingly, the screen readers were more likely to interpret the passage as highlighting God's faithfulness over time than the print readers were. This may indicate an association of print with a more judgmental idea of God and modern technology with a friendlier image of God. It also seems to correspond to Phillips's argument that, in digital environments, users are more likely to share verses that align with Christian Smith's moralistic therapeutic deism.[20] However, the numbers were small enough that this trend cannot be considered definitive without further study.

Finally, the sixth question of the assessment focused not on facts or interpretation, but on how the act of reading influenced the reader's mood. The question asked: "How do you feel after reading Jude (discouraged, encouraged, confused, joyful, etc.)?" Common words among the responses were tallied and, unsurprisingly, many of them contained words from the prompt. However, while many others circled or wrote the words encouraged, discouraged, or confused, only two participants (one print reader and one phone reader) used the word "joyful" in their response. After tallying the results, the most common response among all respondents was "encouraged," but this response was more common for print readers than digital readers (44.1% print, 36.7% digital) (Table 7.5). Conversely, digital users were nearly twice as likely to report feeling "confused" (16.1% print, 30.3% digital). This difference over "confused" had the strongest statistical difference, indicating that there is a strong variance in affective experience between print and digital reading. This affective difference may also contribute to the overall reduced comprehension rates for screen readers.

Though I did not ask the participants if they had read or studied Jude before, many respondents said that they would like to "read it again" or "study Jude more," indicating their relative unfamiliarity with the epistle. All participants, in both the print and screen groups, seemed to feel uplifted by the overall feel of Jude ("encouraged"), and yet they admitted that they did

Table 7.5 Print versus Smartphone Affective Comparison

	Print	Smartphone
Encouraged	44.1%	36.7%
Confused	16.1%	30.3%
Discouraged	7.4%	3.8%

not understand it after a single reading. In the focus group discussion afterward, some indicated that they felt they rarely understood a passage on the first reading and that their usual practice was to read and reread the text several times with study material in hand in order to understand it:

> My favorite part about having it on a device is the vast information available because I don't understand 90% of what I read, even print or not. I've got to study it up.

Others, while reflecting on their reading habits, admitted that they tended to skim more often when reading on a screen compared to print even when they were reading the Bible. This tendency to skim the biblical text on a phone appears to impede the ability to accurately recall the basics of the text on their first reading, especially for male readers. Further, even if they do correctly recall the basic factual elements of the text, the screen appears to induce a mood that is more confused, less spiritually nourished, and yet more attuned to a friendly God than a judgmental one.

If this serves as an example of where the digital Bible software may be affecting evangelical readers in ways they would consider negative or in opposition to their goal of knowing and understanding the Bible, the daily reading plan data below may indicate that Bible software also produces positive effects in line with evangelical values.

Technologically Enhanced Daily Bible Reading

Digital Bible apps, especially YouVersion, place a great deal of emphasis on daily Bible reading, regularly urging users to sign up for plans. But do these plans work, and how do thy compare to a printed plan? To look into this, I asked the participants to engage in a ten-day reading plan for the Gospel of John. For each day of the plan, the participants would read two chapters, or an average of 1,950 words. Participants using smartphones were given a step-by-step instruction sheet showing how to enable the YouVersion plan on their phone and given time to follow the steps and ask questions. During this process, they were also asked to enable reminders and select a time each day for the reminder to appear on their respective smartphones. After ten days, the participants were sent an email with a link to a survey where they could report their reading progress.

Table 7.6 Print versus Smartphone Bible-Reading Plan Comparison by Gender

	Print	Smartphone	*p*-Value
Female	2.83 days	3.77 days	.264
Male	4.58 days	7.75 days	.032
Overall	3.68 days	5.53 days	.053

Table 7.7 Male versus Female Bible-Reading Plan Performance by Medium

	Male	Female	*p*-Value
Print	4.58 days	2.83 days	.140
Smartphone	7.75 days	3.77 days	.032
Overall	6.27 days	3.36 days	.016

It turn out that smartphone users completed a higher percentage of days than print readers (Table 7.6), and, as with the comprehension results, this difference was more acute among male participants than female participants (Table 7.7).

Participants who used smartphones finished almost two more days of reading than print readers, with a high level of statistical significance (print: 3.68 days, phone: 5.53 days). When the gender of the participants was taken in account, the data indicate that women read almost one day more on phones than women read in print, but this difference was not statistically significant (print: 2.83 days, phone: 3.77 days). In contrast, men using a Bible app read an average of more than three additional days compared to those using printed Bibles, and this is a statistically significant result (print: 4.58 days, phone: 7.75 days). This means that, like the difference in comprehension, the overall difference between print and smartphone daily reading can largely be attributed to male readers.

Interestingly, male participants reported reading more days than female participants both in print and on phones. Overall, males finished almost twice as many days as females (male: 6.27 days, female: 3.36 days, $p = .016$). However, this difference was wider on smartphones (male: 7.75 days, female: 3.77) than in printed Bibles (male: 4.58, female: 2.83 days). The printed Bible difference of approximately one and a half days was not statistically significant ($p = .14$), but the smartphone gap of nearly three additional days was statistically significant ($p = .032$), meaning that most of the increase in days between men and women was due to the smartphone gap.

It is noteworthy that the participants used the same medium for both the comprehension assessment above and the reading plan, meaning that the men who scored poorly in reading comprehension using a smartphone were the same group of men who read more days on average using their smartphones. Likewise, the print users who scored higher on comprehension using print were the same group who read fewer days of the reading plan in print. This result is initially surprising because it runs counter to studies indicating that women read the Bible more often than men, and therefore it is possible that results are due to errors in reporting similar to errors seen in church attendance.[21] However, the findings are in line with studies indicating differences between the ways men and women read the Bible and in how they use digital media, which we will consider below.[22]

Smartphone users were also asked if the reminders were helpful in prompting them to read. A small number of participants reported ignoring the reminders and not reading, and others reported that they had not been set up properly and therefore had no effect. But those who did have daily reminders enabled answered that they were helpful, and the data show that these participants read the greatest number of average days of any group. This corresponds with research indicating the digital Bible readers say that reminders "keep me accountable" and help them meet their reading goals.[23] This indicates that the effects of the digital Bible on evangelical readers do not come exclusively from Bible software or phone hardware, but from the combination of the two working together. It also serves as another example of evangelicals who are motivated by Hopeful Entrepreneurial Pragmatism (HEP) adopting existing features (notifications) and repurposing them according to their values.

When Hopeful Becomes Realistic

One important caveat to evangelicals' generally hopeful outlook on technology is that their positive perception of smartphones changed from the beginning of the study to the end. In the initial in-person survey, participants were asked if the digital Bible had had a positive, negative, or mixed effect on their Bible engagement, and they were asked a similar question in the follow-up reading plan survey. In the initial survey of all participants, more than 52% of the participants left comments indicating that digital media had some positive impact on their Bible engagement, 43% said that smartphones had no effect on their religious practices regarding scripture, and only 3% said it had a negative or mixed impact. Several included comments, such as "I'm vastly more engaged since I can read the Bible just a few seconds at a time, anywhere," and many others emphasized how the convenience and accessibility of the phone allowed them to read the Bible more often and in more situations.

This positivity was even more pronounced in those who reported back for the reading plan. In the initial survey, more of them reported a positive impact from digital media (64% compared to the 50% of the wider group) and they were less likely to report negative or mixed feelings (2.5%) (Table 7.8). However, these results shifted significantly after they experienced the reading plan. In the same participants, negative perceptions of phones rose from 0% to 25% while positive perceptions fell dramatically from 64% to 27%, and the number of participants who said digital media had no effect dropped from 34% to 24%.

This shift may indicate that evangelicals tend to have an overall positive impression of technology especially when asked in a group setting, but after actually using it, they are more prepared to acknowledge its downsides and how it may be negatively affecting their spiritual goals. In other words, by default and in group settings, their orientation toward technology is characterized by HEP, but when confronted with specific use cases and offered the chance to respond individually without group dynamics, they are more reflective about technology's trade-offs and downsides.

This may also indicate that Bible software can be considered persuasive, but that is heavily influenced by the hardware on which it runs, and this, in turn, is tempered by a person's current attitude toward phones and their

Table 7.8 Impact of Digital Media

	All Participants	Reading Plan Before	Reading Plan After
Positive	52%	64%	27%
Negative	<1%	0%	25%
Mixed	3%	2.5%	24%
No effect or blank	43%	34%	24%

place in one's individual and social life. At the same time, however, even when evangelicals return to print Bibles, their use of Bible software appears to have reoriented some of their expectations, such as always having access to Bible search and comparison features on their phone and being able to tap a word or verse to see explanatory notes that provide "secondary perspicuity."

Focusing on the Bible in a Digital World

One reader responded to the Bible-reading plan survey with the following comment: "When reading on the phone I feel like I was less engaged than when reading a paper Bible. It felt a little more like skimming an email to get it done rather than really studying God's word. I do like the electronic reminder." Meanwhile, another wrote, "I would have been more consistent if I had used my phone." These comments encapsulate much of the complexity latent in these findings, including that evangelicals value "really studying God's word" and being "consistent," but that there are trade-offs with each medium they use to engage the Bible. They might view the printed Bible as more authentic, but simultaneously see the digital Bible as better at keeping them on track. And they might appreciate the phone reminding them to read, but then feel frustrated by their tendency to skim.

These comments are consistent with the data presented above, indicating that, when it comes to helping evangelicals accomplish their goal of regularly reading the Bible, smartphones are effective, especially when users enabled the reminder feature. However, many users expressed worries that the convenience of the smartphone might be overshadowed by its capacity for distraction and the impulse to skim rather than read deeply, and the data on comprehension indicate that these worries are not unfounded. The paradox of being helpful in one sense and yet worrisome in others was most clearly seen in the results indicating that smartphone users were more likely to read the Bible daily and yet their comprehension suffered on the same medium. Though comprehension cannot be considered the only important element of Bible engagement, it is at least partially a prerequisite for Bible study, and these data would suggest that exclusively using a phone for Bible study may be less than ideal. This leads us to ask what might be causing this lack of comprehension, and why it appears to affect men more than women. Similarly, for evangelicals who value daily reading, we might also ask why the men in this study completed more days of the plan and why they responded to the phone prompts at higher rates than the female participants.

Gender-Based Differences

The gender-based differences observed here were unexpected, but they may be explained by drawing on research from several areas, including children's reading comprehension, gender-based media preferences, and the religious habits of adults. While the research on print and digital media cited above did not find any gender-based comprehension differences in adults, there are indications of gender-based comprehension differences among children. When it comes to reading comprehension tests in a print environment, researchers have found that "girls consistently outperform boys on tests of reading comprehension," but they add that "the reason for this is not clear."[24] Another recent study found that, although boys scored higher on standardized tests, when it comes to reading "girls showed significantly higher reading scores than boys across every wave of assessment and in every grade."[25] The authors do not offer a firm conclusion as to the causes of these differences, but they explore potential reasons including differential rates of maturation, gender differences in lateralization of brain function, gender differences in variability (including boys being overrepresented in populations with reading impairment, dyslexia, attention disorders), gender differences in externalizing behavior and language competence, and gender-stereotyping of reading and language as feminine traits. In one of the very few studies that consider gender when studying screen and print comprehension rates, researchers found a gender-based difference in reading comprehension for print readers among middle-aged adults.[26] However, their data indicated no statistically significant difference in comprehension between male and female subjects for screen reading. In the Bible-reading data above, however, male participants performed better than female participants in print, but females outperformed males when reading on screens.

This comprehension difference appears to be a unique finding among comparisons of print and screen reading, but the gender-based difference in daily Bible-reading habits may be more in line with data suggesting differences in religious behavior and media preference among men and women. Research suggests that women are more religious than men across a variety of cultures, but that technology and secularization tend to reduce the differences in men and women's roles, and it may also be closing the gap in religious behavior.[27] Still, Pew found that women read scripture at higher rates (40% vs. 30% for weekly reading) than men.[28] The American Bible Society's *State of the Bible* also indicates that women read the Bible more often than men, and are more likely to have increased their reading in year over year.

ABS also found that women were more likely to experience a sense of connection with God when reading the Bible. Women were also more likely to install a Bible app and use it regularly, while men were more likely to use an audio version of the Bible.[29] This overall portrait of women being more engaged with the Bible and their faith might explain their higher rates of comprehension, but it might also suggest that they would complete more of the reading plan. However, in Hutchings's survey on digital Bible reading, men claim to read the Bible more often than women, and they also reported being more likely to use Bible app features like digital commentaries and reading plans.[30]

The difference Hutchings found related to Bible-related app usage corresponds with broader data suggesting that men and women have different technology preferences and tend to view their devices differently. As previously mentioned, data suggest a variety of gender-based differences in technology usage, including overall reading frequency, preferred device types, and the kinds of activities men and women prefer to perform on those devices. Confidence levels regarding digital media differ by gender, with women being more likely to believe that all genders have equal ability with computers, but less confidence in their own ability.[31] Other studies indicate that men are more likely to choose competitive environments and overestimate their abilities, and it is possible that overconfidence in their ability to use screens contributed negatively to their comprehension score in the present study, while their competitiveness led them to perform their daily reading more often on both phones and in print.[32]

Studies on leisure habits also suggest additional differences in male and female reading patterns and screen usage. Even before adolescence, young boys and girls show distinct media preferences and reading behavior. A survey from the National Literacy Trust found that boys were more likely to read on screen (65.7%) than in print (55.4%) outside of school.[33] The survey also found that girls were more likely to read print outside of school than boys (boys 55.4%, girls 68.3%). Other researchers have found similar differences in adult reading habits and preferences, such as choosing authors of their own gender and diverging in how men and women go about finding books.[34] In adulthood, a study from the US Bureau of Labor Statistics indicates that men are much more likely to fill their leisure time with "playing games and computer use" and "watching TV," while women were more likely to spend their time "reading" and "socializing and communicating."[35] These data indicate that into adulthood, women continue to be more likely to read for pleasure, while men are more likely to use screens for entertainment during their leisure time.

Traditional gender roles and differences in working hours may also contribute to differences in the net amount of screen usage between men and women. The US Bureau of Labor Statistics indicates that, of men and women who work, men work on average almost a full hour per day more and that women are more likely to work part-time. It is possible that the longer work hours correlate with more digital media usage, which influences screen reading behavior among men. The evangelical participants in this study were not asked directly about income, jobs, or gender roles. However, evangelicals tend to follow "traditional" gender roles and comments made by female participants about reading the Bible while waiting in a carpool to pick up their children from school indicate that the participants in this study may largely follow the traditional division of labor. If they are in line with national statistics on work, the male participants in this study would be more likely to engage in work-related screen reading than the female participants. This increased use of screens during leisure and work and may have influenced the male participants both in their propensity to skim on screen, resulting in decreased comprehension, and in the likelihood that they would respond to phone reminders leading to higher rates of daily Bible reading.

Together, these data indicate that while the comprehension difference found here in Bible reading is unique among print and screen reading tests, it fits within a larger body of research showing gender differences in religious, reading, and media behaviors. Likewise, though ABS's *State of the Bible* indicates that women are more likely to use regularly use Bible apps, the finding here that men completed more of their reading plan using digital media may correspond with gender-based differences in work and leisure usage of screens. Continued work on gender and other demographic categorizations may prove to be a fruitful area for future research as it brings together biblical literacy, technology usage, and religious behavior.

Benefits versus Distractions

Another concern raised by the participants when discussing the digital Bible usage was the problems that they had observed in their own lives with smartphone overuse. When they spoke about their preferred medium for engaging the scriptures, a common refrain was that Bible software itself was not inherently problematic, but because it sits alongside distracting tools that constantly vie for their attention, they found it challenging to keep themselves focused on the Bible. Many of the participants merged technical and religious language, referring to "notifications," "distractions," and "temptations" when speaking of using their phone to read the Bible. The following two participants offered representative comments on the experience of attempting to read but being pulled out of the biblical text by another app:

> For me, anything where it's going to be longer than just a little bit of reference, it's just too easy to get distracted on the phone. . . . There's no long reading there. There's no focus, there's no meditation, because as soon as I get an email, someone likes a tweet, notifies me of my baseball team's score, I'm instantly pulled away from whatever else I was trying to do.
>
> The other reason I don't like digital is because I feel like you get a quick notification of anything on your phone, and all of a sudden I'm not studying what I was studying any more. Now I'm looking at Facebook, or now I'm looking at that text, or now whatever. If I'm reading the [printed] Bible, I can just put this away, and I'm not distracted—other than by kids and stuff, but it's not the same.

Notice that the first user still prefers to use digital tools for "a little bit of reference," but experience has shown her that anything longer leads to distraction. She used words like "focus" and "meditation" to describe the experience she desired from the Bible and noted that the other functions and applications on a phone made these states of mind difficult to achieve. The second reader echoes these comments, focusing on the notifications feature of phones, which come often enough to draw the person's attention away from reading the Bible. He acknowledges that there can be distractions ("kids and stuff") when reading a printed Bible, but sees this as a distinct category because those interruptions are external to the medium of print and perhaps easier to control, whereas the distractions on the phone are an inherent part of its functionality and seem more difficult to manage.

The multipurpose nature of phones led many of the participants to report that they found it challenging to use their phone for a single task for very long. These participants wanted to read or study the Bible, but they found themselves powerless to resist the "quick notification" feature of phones, which leads to being "distracted." But conversely, it was these very notifications that led some of the participants to read the Bible more frequently than print readers or digital users who did not enable notifications. Other participants added to concerns about notifications, worrying that the sheer number of competing apps on a phone might call into question the idea that an electronic Bible would lead to more reading. One participant said, "We've got so much on our phone that we can't keep our focus on anything." These comments seem to be an expression of "attention economics," a field pioneered by Davenport and Beck, who define attention as "focused mental engagement on a particular item of information."[36] Almost a decade before the emergence of the smartphone, they argued that attention would become a more important currency than dollars in the economy to come.

Their ideas were prescient, and resources have emerged that attempt to help people avoid overusing their smartphones. Apple, for example, released a feature called Screen Time in 2018 that shows users how often they use apps and lets them set limits to help them curb activities they would rather not do. This feature was partially in response to the work of former Google design ethicist Tristan Harris, who formed the organization Time Well Spent, which seeks to reverse what he calls the digital attention crisis.[37] In addition, Christian-themed books such as *The Common Rule: Habits of Purpose for an Age of Distraction* have emerged offering guidance on how to structure one's life in such a way as to avoid the pull of digital distractions.[38] This all suggests that Bible software cannot be understood as an app in isolation, but as one of many competitors for the attention of evangelical technology users.

Other users expressed concern about a larger problem of technology dependence that was incompatible with their spiritual values. One participant offered a negative assessment of using a phone to engage the Bible because it fit within a pattern of people becoming too dependent on devices to perform functions they should be able to do themselves. He drew a parallel to the way calculators have led to a decreased ability to do basic mathematical operations:

> Take a look at what the calculator has done to the way we discipline our minds to remember and take in knowledge and data and be able to critically think through that. You take a look at similar things like Google. If I don't know it, what I do? I quickly grab a device and look up what I don't know. . . . Along those lines, same question similar to the calculators, we set up applications that remind us to study [the Bible]. The question I've got is, what does that say about our heart for wanting and desiring to do these

> things? And setting aside and making it important in our own heart in our life to do these things?

Indeed, there are studies indicating that our memories can be weakened by dependence on technology,[39] and Pew data show that smartphone usage continues to rise among all age groups, suggesting that such concerns are not entirely unwarranted.[40] Popular articles regularly warn readers about the danger of technology dependence,[41] but this user seems to be suggesting that while there is no great moral concern in ceding mathematical authority to a calculator or basic knowledge to a Google search, there is a spiritual concern in outsourcing the discipline of a daily Bible to an electronic reminder. Again, the importance of daily Bible reading is not in question, but for this man, the means of accomplishing this goal and consequences of those means should be interrogated. Unlike other participants, he makes a direct connection between the state of one's "heart" and one's use of technology and media. In other words, if a person reads the Bible every day, but only does so because a device reminded him or her, is this spiritual progress or not? The quantitative data from this study suggest that electronic reminders do increase the overall amount and frequency of reading, but as this user suggests, it may not have the same outcome as self-willed reading.

While this participant's comments are framed in terms of the negative qualities of electronic media, other users spoke of how their experience using digital Bibles had created an awareness of positive attributes in printed Bibles that they had not previously named. While the daily reading plans, power searching, and cross-referencing that digital media provide are important and valuable, the following user offered comments about a kind of serendipitous reading that he finds more readily available in print and which he found captures his attention for longer:

> I just find that when I look at the actual physical Bible I'm more apt to stay in there and roam, read the two verses that somebody's mentioned, and then I may continue on. The next thing you know, I've read three pages, where this device, I tend to read just the one or two. I might go one or two verses more, and then I'm out.

Reading a printed book, he feels he is in a "big, open space," while digital media, particularly a phone, feel like "sitting in a little chair." Again, the thrust of his comment is related to the overall evangelical value that more

Bible reading is better, and whatever medium is most capable of producing this is considered superior. But he is arguing that, for all the power evangelical developers have put into the software, ultimately the medium of a print book does more to capture and hold his attention than is possible in a screen-based reading environment. And yet, while many focus group participants expressed similar concerns about attention, focus, and distraction, other users gave examples of how they had found ways to minimize distracting notifications, creating a similar level of focus that they could achieve in reading a printed Bible.

> About all the distractions—I use my Kindle because I don't get notifications. [And I adjusted my phone so that] I don't get many notifications. I've done probably 70 to 80% of my Bible reading this year on my phone, and another 10 or 20% on my Kindle, and precisely 0% paper Bible.

By this participant's estimates, he no longer uses a printed copy of the Bible to engage with scripture, because he intentionally uses either a device without notifications ("my Kindle") or a phone setup with minimal notifications. Interestingly, this is one of the recommended strategies to help people achieve a better life balance. But note that his comments came before the advent of those resources, indicating that evangelicals' willingness to hopefully embrace technology and entrepreneurially experiment with it can sometimes result in a forward-looking relationship with technology that anticipates changes in society rather than merely reacting to them.

The Social Shaping of Bible Software's Impact

In the previous chapter, we saw that Bible software augments the forms of engagement available to Bible readers, and in this chapter, we have observed some effects this change has brought. It appears that evangelical developers have been at least partially successful in changing the behavior of evangelical readers, as demonstrated in the case of daily Bible reading. However, the effects of this change appear to be as much due to the hardware of the phone and its constant presence as it is with the design of the software itself. As YouVersion adds features like "Streaks" (counting how many days or weeks in a row one has read the Bible), smartphones and screens continue to become more deeply integrated into our lives. At the same time, the negative effects on comprehension appear to be caused both by the hardware of the screen and by the larger social shaping context of apps, notifications, and social media.

Audio engineer Damon Krukowski argues that when we as a society adopted the mobile phone, we made a trade-off, sacrificing sound quality in conversation for the convenience of being able to talk to one another anywhere.[42] What we have seen in this chapter is that Bible software presents a similar trade-off between quality Bible reading and the convenience of being read the Bible anywhere. You cannot have a "quiet time" in a grocery store, but you can finish a reading plan. At the same time, Bible software, when used alongside a printed Bible, appears to deepen one's understanding of the text, its interrelatedness, and its importance in one's life. In addition, print, screen, and audio Bibles offer unique ways of enriching relational connections and understanding one's religious identity. In this sense, the digital Bible sits within a much larger network of people, ideas, and rituals, each of which Christians must navigate in their religious experiences.

8
Conclusions

In late 2016, scholars from the University of Kentucky and Hebrew University of Jerusalem announced that they had used X-ray scanners and a tool technique they called "virtual unwrapping" to read the oldest Hebrew scroll ever discovered. The scroll had been part of the Dead Sea Scrolls collection found nearly fifty years ago, but it was damaged in a fire and was charred too badly to unroll. Thanks to their ingenuity in repurposing 3D imaging software that had been funded by Google and the US National Science Foundation, they were able to generate a readable copy of a passage from Leviticus.[1] In a sermon at an evangelical church, this story could be used as a metaphor to illustrate any number of spiritual principles, but here it serves as a reminder that Christianity, the Bible, and technology have had a long, intertwined relationship. From the early Christians who decided to adopt the codex over scrolls, to the invention of the printing press, and to the first attempts at using computers to study the biblical text, there have always been enterprising individuals willing to combine their love of the Bible and their fascination with new technology. However, their experiments did not merely produce new, more useful tools. Instead, those tools and the system of values embedded in them also created new ways of engaging with and understanding the Bible.

This book set out to investigate how Bible software changes the way that readers engage the scriptures and what role evangelical programmers have had in that change. I have taken a social shaping of technology approach in order to examine both the production and consumption sides of Bible software, keeping the broader societal context of technological development in mind. As I have shown, evangelical institutions and individual evangelical technologists have played significant roles in the creation of consumer Bible software, and as evangelical readers have adopted desktop, internet, and mobile software, they have in turn changed and modified some of their reading habits and attitudes about the Bible. In tracing this journey, I have surfaced several key attributes of evangelicalism, including an orientation toward technology theorized as Hopeful Entrepreneurial Pragmatism (HEP). This framework helps explain why evangelical programmers were so attracted to

People of the Screen. John Dyer, Oxford University Press. © Oxford University Press 2023.
DOI: 10.1093/oso/9780197636350.003.0008

creating Bible software and why Christians from evangelical backgrounds have been so willing to embrace Bible apps. At the same time, HEP allows Bible software to serve as an avenue through which to re-examine the way evangelicals tend to negotiate cultural change. Evangelicals often want to engage with popular culture while also resisting it, and Bible software allows them to hold these two impulses in tension. Their hopeful outlook on technology allows them to address moral concerns about new media while also being free to adopt technology as a way of expressing their values. Their entrepreneurial spirit allows them to stake out conservative moral and theological stances while remaining relatively flexible in how they express those convictions. These decisions are evaluated through a pragmatic outlook, an approach that focuses on outcomes and getting to work on a problem even if all the details have yet to be worked out.

In the case of the programmers, we have seen that the history of Bible software is not that of competing industry titans, but an interwoven tapestry of evangelical ministries, universities, and companies. Though some do indeed compete for customers, and many companies have since come and gone, the teams at Faithlife, Bible Gateway, and YouVersion illustrate the flexibility and resilience of evangelicals, especially those with the right mix of technological innovation, business savvy, and good timing. These three companies have very different business models (sales, advertisements, and donations), and they arose in different eras of technology (desktop computers, the internet, mobile devices), but each has been able to successfully merge technological opportunity with their evangelical identity and purpose. The individual programmers and designers within each organization feel that technology is a good gift from God and that part of their Christian calling is to use their skills to benefit the church and reach the world. The HEP orientation allows them to move seamlessly between the realities of running a software company and their beliefs about the mission of God in the world. One important aspect of their mission centers around the shared evangelical emphasis on the importance of the Bible, not merely the belief that it is inspired and authoritative, but that Christians should regularly and substantively engage with it. Though each company goes about it in unique ways, they share a common goal of increasing Bible engagement and a common approach of using any technique they can imagine.

The team at Faithlife tends to focus on increasing the quality of engagement, gearing the content and functionality of Logos toward helping scholars and pastors work smarter and faster. They spoke of seeking out inspiration from a variety of nonreligious technology and design sources in hopes of adopting anything that would make their product better for their customers. YouVersion on the other hand, as a more consumer-oriented product based in a church, has focused its product on increasing the quantity of Bible engagement, following a deep belief that the more people engage with scripture, the more their lives will change. At times, its developers follow trends in the app industry on things like menu navigation and reminders, and at other times, they follow their theological instincts, creating social features that embody their understanding of Christian community. More recently, they have added prayer functionality and additional tools for sharing scripture on social media. In each case, the source of the inspiration is not as important as the end goal: increased Bible engagement for the purpose of changing lives, urging users to more closely follow the way of Jesus. But this mission extends beyond changing the lives of those who are already Christians. Instead, Bible Gateway and YouVersion also want to reach outside the walls of the church and increase Bible engagement globally by bringing more people to faith. At this intersection of Bible and technology and faith and business, each company sees opportunities for fruitful Christian mission, and they continue to work hard to build products that they hope will change the way people read the Bible.

Indeed, this book has also shown that the fruits of their labor have had an influence on the way that Christians interact with the Bible. Hearing a scroll read by a priest in a temple is a quite different experience than reading the Bible on a screen through an app on a bus. But where the codex replaced the scroll, and the printed Bible replaced handwritten ones, Bible software does not appear to be replacing the printed text for everyone. Much like the promise of a paperless office never materialized, it appears that a paperless Christianity is nowhere near the horizon. Instead, while a small group of users reported reading the Bible exclusively in digital form, the majority of participants indicated that they used both print and digital Bibles, sometimes side by side and sometimes choosing one over the other based on the form of engagement they wanted to accomplish. Though they still preferred print for quiet, devotional readings, they like to use their phones for quick searches and their desktops for more in-depth study, and they were divided on other features like note-taking and highlighting. They also indicated that their choice of media was informed and shaped by different social situations. Some users who wanted to read a Bible passage at work chose to use their desktop computer in hopes of being discreet, and even the most tech-savvy parents preferred to read to their children in print. Others preferred to use their phone but refrained from doing so when they felt it might distract others in a worship setting. They also admitted that even though they felt print was better for deep, focused, distraction-free reading, they often chose what I refer to as the Nearest Available Bible (NAB), or the most convenient form of Bible media at their disposal. In most cases today, this is the phone, meaning that the Bible is more available than ever to read, but also that it sits alongside an increasing list of apps vying for one's attention. Bible readers are not simply choosing between print and "Bible software," but instead are making a series of complex choices based on the capabilities of the software, the convenience of the hardware, and the particular social situation. Thus, I have argued that the influence of Bible software cannot be adequately understood apart from the hardware on which it runs and the elements of digital culture that surround and permeate it.

The adoption of digital Bibles and the process of choosing which medium to use for a set of goals and constraints have, in turn, begun to affect how evangelicals engage with and understand scripture. Many Bible apps provide study resources such as notes, commentaries, and original language tools, and the users in this study expressed appreciation for how these helped them in their study. But these features also seem to have shifted their expectations of their print Bibles and even their relationship to the Bible itself. Several users spoke of how the experience of tapping on words and following links to resources within their Bible apps had created a habit such that they expected to have the same experience when they went back to their printed Bible. Perhaps more significantly, some also spoke of relying on these resources rather than "the Bible alone" for interpreting the text. Whereas the Protestant doctrine of perspicuity teaches that meaning of a biblical text can be known by anyone without an authoritative interpreter, what I have called "secondary perspicuity" appears to be an emerging belief among some digital Bible users that they can still arrive at the correct answer as long as they have the right app installed with the right resources unlocked.

In addition to changing attitudes about the nature of the Bible, we have seen that evangelical programmers have been at least partially successful in their hopes of influencing Bible engagement. On the one hand, the data in this study indicate that digital reading plans, such as those offered by YouVersion and other vendors, do tend to increase reading frequency. Although the long-term effect of these plans on reading is unclear, in the short term, by adopting conventions like phone notifications and taking advantage of the social nature of phones and digital media, YouVersion appears to be meeting its goal of encouraging people to read the Bible more regularly. At the same time, this study brings up questions about what comes alongside increased screen-based Bible reading. At the beginning of the study, many of the participants reported a very positive outlook toward technology, but after the comprehension assessment and ten-day reading plan, a large percentage changed their view. Initially, they expressed a more neutral outlook on technology indicative of HEP, but after an experience with technology and time to reflect on it, many were less enthusiastic and more likely to speak of Bible apps in terms of trade-offs. One of these trade-offs came in the comprehension assessments, which found that reading a Bible on a screen tended to decrease comprehension, especially for men. In addition, screen readers were less likely to say they had experienced a sense of "encouragement" and more likely to report being "confused." Granted, they were reading Jude, which is not among the Bible's most straightforward or encouraging texts, but whatever confusion Jude might produce on its own, the effects seemed to be more pronounced in the digital environment. At the same time, when asked about the meaning of Jude, the phone readers were more likely to speak of God in terms of faithfulness, while print readers emphasized the judgments in the passage.

These observations point to areas of future research that could build upon the work presented here. The longitudinal data on Bible app adoption gathered by the American Bible Society continue to be very helpful in tracking overall electronic Bible usage relative to print, and this thesis adds color and meaning to those data by looking at specific forms of engagement, comprehension, and daily reading. In the future, more work could be done to understand how screen-based Bibles influence the way people conceive of the Bible itself and how they interpret it. In the example above, this study hinted that reading the Bible on a screen may cause some readers to see a kinder, gentler God instead of the judgmental God they might see in print. Future researchers could replicate the print and screen comparison studies approach modeled here, but broaden them to include more types of biblical genres, including poetry, prophecy, and narrative. And, in addition to assessing comprehension and interpretation, future studies could also develop additional metrics that would further probe how media influence the experience of Bible engagement. Such studies could also be conducted among larger, more diverse populations of readers, including those from different denominational and ethnic backgrounds, providing a richer data set from which to examine how different Christian traditions understand the relationship between technology and the Bible. The growing use of audio Bibles also appears to be a relatively untouched avenue for research, and one that could draw from the deep well of research on orality and postliteracy. Finally, expanding this study outside of the West and into the Global South and East where Christianity is growing might shed light on where Christianity and the digital Bible are headed.

The Digital Bible in Modern Faith

Much of the material in this book came from my PhD dissertation for Durham University. During my Viva defense, one of the examiners noted that I had not done much theologizing about the digital Bible and the faithful Christian life, and he asked why. I replied that, as an evangelical myself, I am aware that my religious heritage has sometimes taken a posture of telling people what to do and why they are wrong, rather than taking the time to listen and understand. For this project, however, I wanted to learn how to step back and use the tools of sociology and digital religion to describe a phenomenon in a way that contributes to the academy. But in choosing to be largely nonprescriptive, I am hopeful that this description will be helpful to fellow Christians navigating their faith in a digital world, especially as the experience of a pandemic lingers.

Based on what I have observed about the production and use of the digital Bible, I might offer the following theologically motivated recommendations to both developers and regular readers. To the creators of Bible software and other religious apps—among whom I count myself—I would first of all say, thank you. Whatever faults we might find or quibbles we might have with (persuasive) design choices, what you have created has enabled us to engage with scripture in fresh ways, to see new things about God, and to connect with others around the shared identity marker of scripture. At the same time, I would encourage us all to carefully consider the unintended ways all media, technology, and culture can shape, form, and inform our work. Exploring the benefits and the dangers of pragmatism and interrogating the differences between the kingdom of God and the empire of humanity should be an essential part of creative, entrepreneurial work. It is an awesome responsibility to make choices that could affect millions of people, and we need to ensure that these choices are driven by conviction, not profit alone. Finally, I would urge us to continue experimenting, drawing on approaches to scripture from a wide range of traditions and approaches to technology from the breadth of the tech industry, and evaluating these works not only for how they function but for what they mean to users and the habits and hermeneutics they teach.

To the users of digital Bibles, I would offer much the same advice. First, we ought to be incredibly grateful for the gift of almost unlimited access to translations, original languages, audio recordings, resources, and tools that the world of the digital makes possible. What multimedia riches we have! And yet, just as developers need to be aware of the power they and their tools have over our reading habits and hermeneutics horizons, we readers need to be aware of the ways in which medium and message are inseparable. The idea that the means can change while the message stays the same may sound good in a tweet, but it does not reflect how humans, technology, and community actually work. The new patterns of Bible engagement available to us—reading scripture on a screen, sharing scripture on social media, searching in a language we do not speak, hearing scripture in the car, and so so—are no more neutral than the advent of the printed Bible several centuries ago. In this digital era, I would encourage you to mix old and new, memorize not just search, meditate not just share, listen not just read, do not just hear. As you use different forms of media to encounter scripture, reflect on them with others in your faith community, and work together to make choices out of conviction rather than convenience alone.

A final story is important for both developers and readers. One of the first questions I asked the participants in this study was how they felt their engagement with the Bible had changed since they first started using Bible apps. Many of them shared stories of their excitement and frustration with digital Bibles, but a few surprised me when they said they regularly used digital Bibles, but it had not brought any change. This initially puzzled me, until I read on responder's explanation: "I became a Christian after the advent of the smartphone." These words remind us that, as a new generation encounters the Bible for the first time, they will not experience it exclusively orally as in the days before the printing press, or primarily in print as was the case for the last several centuries. Instead, for them, "the Bible" will always be a multimedia category, and they will have more complex decisions to make about which combination of Bible media they want to use. If the evangelicals in this study are any indication, the next generation will continue to find ways to faithfully navigate whatever comes, embracing new technology while holding onto what they believe is essential. As the prophet Isaiah might say, technology will advance and media will change, but the scriptures will remain forever.

APPENDIX

Bible Software List

Desktop Software, 1982–1987

Table A.1 is a list of several dozen Bible software applications, including the date the app was launched (where available) and the name and location of the company behind it, compiled from John Hughes's *Bits, Bytes and Biblical Studies* (1987) unless otherwise indicated. Entries marked with an asterisk (*) are derived from a blog post by the current author where Bible software developers added details not present in Hughes's work.[1] Those without a clear date are marked "<1987" to indicate that they were released before the publication of Hughes's book.

Table A.1 Bible Software, 1982–1987

Year	Name and Company	Location
1982	*The Word Processor*, Bible Research Systems	Austin, TX
1984	*Scripture Scanner*, Omega Software	Round Rock, TX
1985	*compuBIBLE*, National Software Systems	Borger, TX
1985	*Bible Search*, Scripture Software	Orlando, FL
<1987	*BIBLE-ON-DISK*, Logos Information Systems	Sunnyvale, CA
1985[a]	*COMPUTER BIBLE*, Computer Bibles International[b]	Greenville, SC
1985	*Bible-Reader*, Philip Kellingley	United Kingdom
1986	*The Powerful Word*, Hatley Computer Services	Springfield, MO
1986	*EveryWord Scripture Study System*, Echo Solutions	Provo, UT
1986	*GodSpeed*, Kingdom Age Software *Greg Eskridge*	Plano, TX
<1987	*Com Word 1*, Word of God Communication	Thousand Oaks, CA
<1987	*Wordworker: The Accelerated New Testament*, The Way International	Knoxville, OH
<1987	*KJV on DIALOG*, DIALOG Information Retrieval Service	Palo Alto, CA
<1987	*COMPUTER NEW TESTAMENT*, The Spiritual Source	Manorville, CA
<1987	*INTERNATIONAL BIBLE SOCIETY TEXT*, The International Bible Society	East Brunswick, NJ
<1987	*VERSE BY VERSE*, G.R.A.P.E., Gospel Research and Program Exchange	Keyport, WA

(continued)

Table A.1 Continued

Year	Name and Company	Location
<1987	*MacBible*, Encycloware	Ayden, NC
<1987	*MacConcord / MacScripture*, Medina Software	Longwood, FL
<1987	*New Testament Concordance*, Midwest Software	Farmington, MI
<1987	*The Lamp*, Special Computers Services*	Berrien Springs, MI
<1987	*EveryWord*, EveryWord*	Orem, UT
<1987	*Bible Soft**	Homedale, ID
1987[c]	*Online Bible*	Canada

[a] A newspaper article from late 1984 indicates it would be released the next year. "Memos," *Courier-Journal* (Louisville, KY), December 17, 1984, https://www.newspapers.com/newspage/109761475/.

[b] Frank D. Larkins, founder of Computer Bibles International, was awarded a patent in 1983 for "Hand held portable computer in the form of a book." "Hand Held Portable Computer in the Form of a Book," updated September 19, 1983, https://patents.justia.com/patent/D284966.

[c] An older version of the Online Bible website says it was started "in 1987, long before there was a world wide web." "Online Bible Homepage," 2018, accessed February 24, 2017, http://web.archive.org/web/20080312151153/http://www.onlinebible.org/.

Desktop Software, 1987–2000

Table A.2 is a list of applications that appeared between 1987 and 2000, compiled from Jeffrey Hsu's *Comprehensive Guide to Computer Bible Study* (1993) unless otherwise indicated. Web-based lists (•) and (▾) have been augmented with information from company websites when available.[2] This table does not include all freeware, shareware, and Unix-based software, but it does include software mentioned in a list or review. When a start date was not available, the year is designated as "<1993" to indicate the article or book in which the software were listed.

Table A.2 Bible Software (1987–2000)

Year	Name and Company	Location
1987	*WORDsearch*, Navpress software	North Carolina Colorado Springs, CO
1988	*PC Study Bible*, BibleSoft	Seattle, WA
1988	*QuickVerse* (Logos Bible Processor), Parsons Technology (Creative Computer Systems)	Hiawatha, IA
1988	*CDWord*, Dallas Theological Seminary	Dallas, TX
1992	*Logos Bible Software* Faithlife (Logos)	Bellingham, WA
1992	*BibleWorks* (Bible Word Plus), BibleWorks, LLC (Hermeneutika)	Big Fork, MO Norfolk, VA
1992[a]	*God's Word for Windows*, Kevin Rintoul•	Victoria, B.C., Canada

Table A.2 Continued

Year	Name and Company	Location
<1993	*Ask God*, Integrated Systems and Information	Kirkland, WA
<1993	*Bible Link*, Eagle Computing	Elizabethton, TN
<1993	*Bible Master*, American Bible Sales	Anaheim, CA
<1993	*Bible Source*, Zondervan Electronic Publishing[*]	Grand Rapids, MI
<1993	*Everyword Good News Bible*, [b] American Bible Society	New York, NY
<1993	*Holy Bible*, Window Book	Cambridge, MA
<1993	*Holy Scriptures*, Christian Technologies	Independence, MO
<1993	*Seedmaster*, White Harvest Software[*]	Raleigh, NC
<1993	*The Word Advanced Study System*, Wordsoft/Word	Irving, TX
<1993	*Thompson Chain Hyperbible*, Kirkbride Software[*]	Indianapolis, IN
<1993	*Master Search Bible*, Tri Star Publishing	Horsham, PA
1994	*Accordance*, Oak Tree Software	Altamonte Springs, FL
1995	*Sword Searcher* (Bible Assistant),[c] StudyLamp Software[*]	Broken Arrow, OK
1995	*Bible Explorer*, Epiphany Software[*]	San Jose, CA
<1997[d]	*Rainbow Study Bible*, Rainbow Studies International[*]	
1997[e]	*Theophilos*, Theophilos Bible Software (Sold to Laridian)[*]	
1998	*The Bible Library*, Ellis Enterprises[f]	Oklahoma, OK
<1998	*iBiblio (Bible Windows)*,[g] Silver Mountain Software	Cedar Hill, TX
<1998	*FindIt*, American Bible Society[*]	New York, NY
<1998	*Bible on Disk for Catholics*, Liguori Faithware[*]	
<1998	*Deluxe Bible for Windows*, Rocky Mountain Laboratories[*]	Fort Collins, CO
<1998	*Douay-Rheims version of Holy Bible on PC*, Preserving Christian Publications	Philadelphia, PA
<1998	*Bible Companion*, White Harvest Software[*]	Raleigh, NC
<1998	*DataBiblen-GodSpeed*, SigveSaker[*]	Stavanger Norge, Norway
<1998	*macBible*, Zondervan[*]	Grand Rapids, MI
<1998	*AnyText*, Linguist's Software[*]	Edmonds, WA
<1998	*BibleBrowser*, HolyMac[*]	
1999	*Power Bible*, Online Publishing (powerbible.com)	Branson, MI

(*continued*)

Table A.2 Continued

Year	Name and Company	Location
2003	*iLumina Bible*, Tyndale Publishing House	
2011	*The Discovery Bible*, Gary Hill, H.E.L.P.S. Ministries	

[a] The earliest record of God's Word for Windows was a version available for Windows 3.1 released in 1992. "Island Code Works Homepage," 1999, accessed February 24, 2017, http://web.archive.org/web/19991127121128/http://islandcodeworks.com/.

[b] This product shares the name "EveryWord" with the company EveryWord, Inc., but it is unclear if they were related.

[c] "The History of SwordSearcher," updated March 2018, https://www.swordsearcher.com/history-of-swordsearcher-bible-software.html.

[d] "Rainbow Studies Homepage," March 27, 1997, https://web.archive.org/web/19970327090752/http://www.rainbowstudies.com/.

[e] "About Us," n.d., accessed June 1, 2019, http://theophilos.com/aboutus.htm.

[f] "A History of Ellis Enterprises," September 19, 2000, https://web.archive.org/web/20000919233657/http://www.biblelibrary.com/company.htm.

[g] "Bible Windows" was created by John Baima and the company he founded in 1982, Silver Mountain Software. Microsoft later claimed the name "Bible Windows" infringed on its copyright for the name "Windows," forcing Baima to change the name to Bibloi.

Mobile Era

Major apps for mobile systems are noted in Table A.3, with their original operating system or mobile platform. By the mid-2010s, the number of mobile apps began to grow exponentially, especially as some companies released a new app for each translation so they could be more easily discovered in the Apple and Android's app stores and purchased individually.

Table A.3 Mobile Apps

Please note that this is not an exhaustive list of mobile apps, but only a record of early, significant apps, and a representative sample of apps released since the iPhone and Android were released.

Year	Application Name and Company	Platform
1996	*The Holy Bible—King James Version*, K2 Consultants	Apple Newton
1997	*The Message for Newton*, Servant Software	Apple Newton
1998	*PocketBible*, Laridian	Pocket CE
1998	*BibleReader*, Drew Haninger (Olive Tree Software)	Palm
1998	*Bible for Gameboy*, Wisdom Tree (www.wisdomtreegames.com)	Gameboy
1998	*BibleReader*, Drew Haninger (Olive Tree Software)	Palm
1999	*BibleReader*, Drew Haninger (Olive Tree Software)	Pocket PC

Table A.3 Continued

Year	Application Name and Company	Platform
2003	*Palm Bible+*, Poetry Poon (palmbibleplus.sourceforge.net)	Palm
2008	*YouVersion* (iPhone), YouVersion / Life.Church, Edmond, OK	iOS
2011	*gloBIBLE* (globible.com), Fishermen Labs	iOS
2014	*She Reads Truth*, Raechel Myers and Amanda Bible Williams	iOS, Android
2015	YouVersion for Kids, Life.Church	iOS, Android
2017	*NeuBible*, Kory Westerhold and Aaron Martin	iOS
2017	*Our Bible, App* Crystal Cheatham	iOS, Android
2018	*Bible.is*, Faith Comes by Hearing	iOS, Android
2020	*Spark Bible App*, Spark Bible	iOS, Android

Notes

Chapter 1

1. Carter's presidential run led several publications to follow George Gallup in declaring 1976 the "year of the evangelical." David Kucharsky, "Year of the Evangelical '76," *Christianity Today* 21, no. 2 (1976); Lyman A. Kellstedt et al., "Religious Voting Blocs in the 1992 Election: The Year of the Evangelical?," *Sociology of Religion* 55, no. 3 (Fall 1994); Jon Meacham, "The Editor's Desk," *Newsweek*, November 12, 2006, http://www.newsweek.com/editors-desk-106637.
2. "25th Anniversary for Bible Software Is Celebrated with New Release," Business Wire, updated October 11, 2005, https://www.businesswire.com/news/home/20051011005141/en/25th-Anniversary-Bible-Software-Celebrated-New-Release; John Hughes, *Bits, Bytes and Biblical Studies* (Grand Rapids, MI: Zondervan, 1987), 346; Scott Lubeck, "Scrolling through the Bible," *Texas Monthly*, December 1984, 202.
3. David Hunter, "How They Got the Bible on Disk," *Softalk Mag*, May 9, 1982, 256.
4. Jeffrey and others mention the possibility that "people of the book" was coined by Muhammad, the Prophet of Islam, in the Quran. David Lyle Jeffrey, *People of the Book: Christian Identity and Literary Culture* (Grand Rapids, MI: Eerdmans, 1996), xi.
5. Roberts, Skeat, and Gamble offered several proposals for why Christians chose to use the codex before it was socially acceptable for important literature. Harry Y. Gamble, *Books and Readers in the Early Church: A History of Early Christian Texts* (New Haven: Yale University Press, 1995); T. C. Skeat, "The Origin of the Christian Codex," *Zeitschrift für Papyrologie und Epigraphik* 102 (1994); Colin H. Roberts and T. C. Skeat, *The Birth of the Codex*, repr. ed. (Oxford: Oxford University Prress, 1987); Roberts and Skeat, *Birth of the Codex*; Colin H. Roberts, *The Codex* (London: British Academy, 1954).
6. Elizabeth L. Eisenstein, *The Printing Press as an Agent of Change: Communications and Cultural Transformations in Early Modern Europe*, 2 vols. (Cambridge: Cambridge University Press, 1979), 163–452.
7. Chapter divisions were created before the printing press, and there were other versification schemes that predated Estienne's, but his system eventually became the most widely adopted, and all subsequent Bibles were standardized on it.
8. Bryan Bibb, "Readers and Their E-Bibles: The Shape and Authority of the Hypertext Canon," in *The Bible in American Life*, ed. Philip Goff, Arthur E. Farnsley II, and Peter J. Thuesen (Oxford: Oxford University Press, 2017).
9. Nicholas Carr, "Is Google Making Us Stupid?," *The Atlantic*, July 2008, https://www.theatlantic.com/magazine/archive/2008/07/is-google-making-us-stupid/306868/.

10. Huang and Lu, for example, track the iterative development of an e-commerce website using hygiene-motivation theory. Travis K. Huang and Fu Fong-Ling, "Understanding User Interface Needs of E-commerce Web Sites," *Behaviour & Information Technology* 28, no. 5 (2009), https://doi.org/10.1080/01449290903121378. Others have argued that mobile software increases user participation in the development process, which we will see when we approach mobile Bible software. Lela Mosemghvdlishvili and Jeroen Jansz, "Negotiability of Technology and Its Limitations," *Information, Communication & Society* 16, no. 10 (2013), https://doi.org/10.1080/1369118X.2012.735252.
11. Jeffrey Siker, *Liquid Scripture: The Bible in a Digital World* (Minneapolis: Fortress Press, 2017), 93.
12. Heidi A. Campbell, *Digital Creatives and the Rethinking of Religious Authority Online* (London: Routledge, 2021).
13. Tim Hutchings, "Design and the Digital Bible: Persuasive Technology and Religious Reading," *Journal of Contemporary Religion* 32, no. 2 (2017), https://doi.org/10.1080/13537903.2017.1298903.
14. Timothy E. W. Gloege, *Guaranteed Pure: The Moody Bible Institute, Business, and the Making of Modern Evangelicalism* (Chapel Hill: University of North Carolina Press, 2015).
15. Bruce Shelley, *Evangelicalism in America* (Grand Rapids, MI: Eerdmans, 1967), 7.
16. This approach to technology is expressed well in the warnings of Brooke. Tal Brooke, *Virtual Gods: The Seduction of Power and Pleasure in Cyberspace* (Eugene, OR: Harvest House, 1997).
17. Nir Eyal, *Hooked: How to Build Habit-Forming Products* (New York: Penguin, 2014), 179–82.
18. Depending on the company, this might take the form of sales, downloads, interactions, page views, or other metrics, defined by the technology and business model.
19. James S. Bielo, *Words upon the Word: An Ethnography of Evangelical Group Bible Study* (New York: New York University Press, 2009); Vincent Crapanzano, *Serving the Word: Literalism in America from the Pulpit to the Bench* (New York: New Press, 2000). The Canadian Bible Forum's study of Bible usage is emblematic of how evangelicals view Bible engagement. Rick Hiemstra, *Confidence Conversation and Community: Bible Engagment in Canada, 2013* (Toronto: Faith Today Publications, 2014), http://files.evangelicalfellowship.ca/research/CBES-Report-Confidence-Conversation-and-Community.pdf.
20. Campbell and her students laid out a framework for religious mobile applications. Heidi A. Campbell et al., "There's a Religious App for That! A Framework for Studying Religious Mobile Applications," *Mobile Media & Communication* 2, no. 2 (2014). Fackler explored the rise of niche Bible products designed for specific markets beginning in the mid-twentieth century driven by the evangelical publishing industry. Mark Fackler, "The Second Coming of Holy Writ: Niche Bibles and the Manufacture of Market Segments," in *New Paradigms for Bible Study: The Bible in the Third Millennium*, ed. Robert M. Fowler, Edith Blumhofer, and Fernando F. Segovia (New York: T & T Clark International, 2004).

21. Bobby Ross Jr., "With 330 Million Downloads, Top Bible App Celebrates 10 Years," *Religion News Service*, July 26, 2018, https://religionnews.com/2018/07/26/with-330-million-downloads-top-bible-app-celebrates-10-years/.
22. Apple, *iTunes Charts* (2018), https://www.apple.com/itunes/charts/free-apps/.
23. American Bible Society, *The State of the Bible 2022*, American Bible Society, April 2022, xii, https://1s712.americanbible.org/state-of-the-bible/stateofthebible/State_of_the_bible-2022.pdf.
24. American Bible Society, *The State of the Bible 2019*, American Bible Society, March 2019, 85, https://1s712.americanbible.org/state-of-the-bible/stateofthebible/State_of_the_bible-2019.pdf.
25. Duane Harbin, "Fiat Lux," in *Formatting the Word of God: An Exhibition at Bridwell Library*, ed. Valerie R. Hotchkiss and Charles C. Ryrie (Dallas, TX: Bridwell Library, 1998).
26. Hughes, *Bits, Bytes and Biblical Studies*.
27. "Dear Pastor, Bring Your Bible to Church," Gospel Coalition, 2013, http://thegospelcoalition.org/article/dear-pastor-bring-your-bible-to-church.
28. "Infographics," 2014, http://blog.youversion.com/infographics/.
29. "Easter Bible Engagement around the World," updated April 6, 2015, http://blog.youversion.com/2015/04/easter-bible-engagement-around-the-world/.
30. "After Mass Shootings, Americans Turn to Four Bible Verses Most," *Christianity Today*, updated October 2, 2017, http://www.christianitytoday.com/news/2017/october/after-mass-shootings-top-bible-verses-psalm-34-18-las-vegas.html.
31. "2020's Most-Read Bible Verse: 'Do Not Fear,'" updated 3 December, 2020, https://www.christianitytoday.com/news/2020/december/most-popular-verse-youversion-app-bible-gateway-fear-covid.html.
32. American Bible Society, *The State of the Bible, 2014–2022*. Available from https://sotb.research.bible/.
33. American Bible Society, *The State of the Bible 2019*, 84–87.
34. American Bible Society, *The State of the Bible 2022*, 20.
35. "3 Takeaways from New Research on Americans and the Bible," updated April 24, 2018, https://www.thegospelcoalition.org/blogs/trevin-wax/3-takeaways-new-research-americans-bible/. This refers to Christian Smith and Melinda Lundquist Denton, "Moralistic Therapeutic Deism," in *Soul Searching: The Religious and Spiritual Lives of American Teenagers* (Oxford: Oxford University Press, 2005), 118.

Chapter 2

1. These ideas were initially laid out in *Understanding Media* and later expanded in other works and by his students. Marshall McLuhan, *Understanding Media: The Extensions of Man* (New York: New American Library, 1964).
2. Marshall McLuhan, *The Gutenberg Galaxy: The Making of Typographic Man* (Toronto: University of Toronto Press, 1962).

3. McBride called McLuhan "the most important—and most influential—thinker Canada has ever produced." Jason McBride, "In the '60s, Marshall McLuhan was Toronto's Most Famous Intellectual; Now, the World Has Finally Caught Up with Him," *Toronto Life* (Toronto), July 6, 2011, https://torontolife.com/city/marshall-mcluhan-profile/.
4. For the history, progression, and significance of media ecology as a discipline, see Lance Strate, *After Television in Trees: Echoes and Reflections on Media Ecology as a Field of Study* (New York: Hampton Press, 2006); *Perspectives on Culture, Technology and Communication: The Media Ecology Tradition*, ed. Casey Man Kong Lum (New York: Hampton Press Communication, 2005); Chris Ridgeway, "Scripture in Digital Context: Explorations in Media Ecology and Theology" (2009), https://www.chrisridgeway.net/thesis.
5. Ellul's two most important works in this area are Jacques Ellul, *The Technological Society* (New York: Vintage, 1964), and *Propaganda: The Formation of Men's Attitudes* (New York: Vintage, 1972).
6. For a collection of essays exploring questions of technological determinism, see *Does Technology Drive History? The Dilemma of Technological Determinism*, ed. Leo Marx (Cambridge, MA: MIT Press, 1994).
7. John Fekete, "McLuhanacy: Counterrevolution in Cultural Theory," *Telos* 15 (Spring 1973): 75.
8. A summary of his thought is John Culkin, "Each Culture Develops Its Own Sense Ratio to Meet the Demands of Its Environment," in *McLuhan Hot and Cool: A Primer for the Understanding of McLuhan*, ed. Gerald Emanuel Stearn (New York: New American Library, 1967), 52. McLuhan himself would deny "inevitability" Marshall McLuhan and Quentin Fiore, *The Medium Is the Massage* (New York: Bantam, 1967), 25.
9. Jacobs offers a helpful assessment of McLuhan, regarding him as both frustrating and necessary reading. Alan Jacobs, "Why Bother with Marshall McLuhan?," *New Atlantis*, Spring 2011; Lance Strate, "President's Message: Understanding MEA," *In Medias Res* 1, no. 1 (1999): 1.
10. Andrew Feenberg, "What Is Philosophy of Technology?," lecture for the Komaba undergraduates, 2003.
11. Social shaping of technology was first laid out by the influential collection *The Social Shaping of Technology*, ed. Donald MacKenzie and Judy Wajcman (Milton Keynes, England: Open University Press, 1985).
12. David Edge, "The Social Shaping of Technology," Edinburgh PICT Working Paper no. 1, Edinburgh University, 1988.
13. Stephen Choi, "Explaining African-American Cell Phone Usage through the Social Shaping of Technology Approach," *Journal of African American Studies* 20 (2016), for example, uses social shaping methodologies to explore differences in technology adoption by race and ethnicity.
14. Janet Fulk, "Social Construction of Communication Technology," *Academy of Management Journal* 36, no. 5 (1993): 922.
15. Wiebe E. Bijker, *Of Bicycles, Bakelites, and Bulbs* (Cambridge, MA: MIT Press, 1995), 18–99.

16. Cynthia Cockburn and Susan Omrud, *Gender and Technology in the Making* (London: Sage Publications, 1993), 41–74.
17. Laura P. Schaposnik and James Unwin, "The Phone Walkers: A Study of Human Dependence on Inactive Mobile Devices," *eprint arXiv* 1804.08753, April 25 2018.
18. Robin Williams and David Edge, "The Social Shaping of Technology," *Research Policy* 25 (1996): 866.
19. Wiebe E. Bijker, Thomas P. Hughes, and Trevor Pinch, "Introduction," in *The Social Construction of Technological Systems: New Directions in the Sociology and History of Technology*, ed. Wiebe E. Bijker, Thomas P. Hughes, and Trevor Pinch (Cambridge, MA: MIT Press, 1987).
20. Clay McShane, "The Centrality of the Horse to the Nineteenth-Century American City," in *The Making of Urban America*, ed. Raymond Mohl (New York: SR Publishers, 1997).
21. The term "Bible engagement" is a general term used to encompass any interaction with the Bible, including memorization, study, and *lectio divina*. Its use in research and among evangelicals will be explored below.
22. Apple has long been noted for its emphasis on simplicity, as noticed in early reviews of its desktop software and more recent design analysis. Cameron Shelley, "The Nature of Simplicity in Apple Design," *Design Journal* 18, no. 3 (2015), https://www.tandfonline.com/action/showCitFormats?doi=10.1080%2F14606925.2015.1059609; Gregg William, "Review: The Apple Macintosh Computer," *Byte*, 1984, 31.
23. Eric Vaniman, "Select Scrolling Options in Your Logos Bible App," *Logos Talk*, 2015, https://web.archive.org/web/20150919235942/https://blog.logos.com/2015/03/select-scrolling-options-in-your-logos-bible-app/.
24. Hans Klein and Daniel Lee Kleinman, "The Social Construction of Technology: Structural Considerations," *Science, Technology, & Human Values* 27, no. 1 (2002): 31.
25. Minjeong Kang and Juhyun Eune, "Design Framework for Multimodal Reading Experience in Cross-Platform Computing Devices—Focus on a Digital Bible," paper presented at the DRS 2012 Bangkok, Chulalongkorn University, Bangkok, 2012.
26. Langdon Winner, "Upon Opening the Black Box and Finding It Empty: Social Constructivism and the Philosophy of Technology," *Science, Technology, and Human Values* 18, no. 3 (1993).
27. Scheifinger, "The Significance of Non-participatory Digital Religion: The Saiva Siddhanta Church and the Development of a Global Hinduism," in *Digital Hinduism: Dharma and Discourse in the Age of New Media*, ed. Murali Balaji (Lanham, MD: Lexington Books, 2017).
28. Klein and Kleinman, "Social Construction of Technology," 30.
29. Bijker, *Bicycles, Bakelites, and Bulbs*, 126.
30. Rheingold documented the early use of bulletin board systems for religious discourse, including a "create your own religion" thread in chapter 2 of his book. Howard Rheingold, *The Virtual Community: Homesteading on the Electronic Frontier* (New York: Harper Perennial, 1993).
31. Lorne Dawson, "Researching Religion in Cyberspace: Issues and Strategies," in *Religion on the Internet: Research Prospects and Promises*, ed. Douglas E. Cowan and Jeffrey K. Hadden (New York: HAI Press, 2000), 29.

32. Brenda Brasher, *Give Me That Online Religion* (San Francisco: Jossey-Bass, 2001), 9.
33. Morten T. Højsgaard and Margit Warburg, "Introduction: Waves of Research," in *Religion and Cyberspace* (New York: Routledge, 2005), 1–12.
34. Heidi A. Campbell and Brian Altenhofen, "Digitizing Research in the Sociology of Religion," in *Digital Methodologies in the Sociology of Religion*, ed. Sariya Cheruvallil-Contractor and Suda Shakkour (New York: Bloomsbury, 2016).
35. Mark Kellner, *God on the Internet* (Hoboken, NJ: John Wiley & Sons, 1996), 242.
36. Stephen D. O'Leary, "Cyberspace as Sacred Space: Communicating Religion on Computer Networks," *Journal of American Academy of Religion* 64, no. 4 (1996): 782.
37. Stephen D. O'Leary and Brenda Brasher, "The Unknown God of the Internet: Religious Communication from the Ancient Agora to the Virtual Forum," in *Philosophical Perspectives on Computer-Mediated Communication*, ed. Charles Ess (Albany: State University of New York Press, 1996), 251.
38. Phil Mullins, "Imagining the Bible in Electronic Culture," *Religion & Education* 23, no. 1 (June 1, 1996): 38, https://doi.org/10.1080/15507394.1996.11000822.
39. Tom Beaudoin, *Virtual Faith: The Irreverent Spiritual Quest of Generation X* (San Francisco: Jossey-Bass, 1998), 127.
40. "How to Build Your Own Bible," Smile Politely, updated August 12, 2009, http://www.smilepolitely.com/opinion/how_to_build_your_own_bible/.
41. Thomas Jefferson et al., *The Jefferson Bible: The Life and Morals of Jesus of Nazareth, Extracted Textually from the Gospels in Greek, Latin, French & English* (Washington, DC: Smithsonian Books, 2011).
42. *Religion on the Internet: Research Promises and Prospects*, ed. Jeffrey K. Hadden and Douglas E. Cowan (Bingley, UK: Emerald Group Publishing, 2000).
43. Christopher Helland, "Online-Religion/Religion-Online and Virtual Communities," in *Religion on the Internet: Research Prospects and Promises*, ed. Douglas E. Cowan and Jeffrey K. Hadden (New York: HAI Press, 2000).
44. Helland, "Online-Religion/Religion-Online," 207.
45. One of Campbell's primary conclusions about recent research on religion and the internet is that the distinction between online and offline has itself blurred, as users no longer think of themselves as "going online" but instead fluidly move between online and offline context throughout the day. Heidi A. Campbell, "Understanding the Relationship between Religion Online and Offline in a Networked Society," *Journal of the American Academy of Religion* 80, no. 1 (2012); Douglas Cowan, *Cyberhenge: Modern Pagans on the Internet* (New York: Routledge, 2004), 19; Glenn Young, "Reading and Praying Online: The Continuity of Religion Online and Online Religion in Internet Christianity," in *Religion Online: Finding Faith on the Internet*, ed. Douglas Cowan and Lorne Dawson (New York: Routledge, 2004).
46. Christopher Helland, "Online Religion as Lived Religion: Methodological Issues in the Study of Religious Participation on the Internet," *Heidelberg Journal of Religions on the Internet* 1, no. 1 (2005): 5.
47. Our Bible App Mission, "Mission: About Our Bible App," 2018, accessed September 2018, https://www.ourbibleapp.com/mission/.

48. Christine Hine, *Virtual Ethnography*, online resource (London: Sage, 2000), 63–65; Robert V. Kozinets, *Netnography: Ethnographic Research in the Age of the Internet* (Thousand Oaks, CA: Sage, 2010), 41ff.
49. Lorne L. Dawson, "Religion and the Quest for Virtual Community," in *Religion Online: Finding Faith on the Internet*, ed. Lorne L. Dawson and Douglas E. Cowan (New York: Routledge, 2005).
50. Tim Hutchings, *Creating Church Online: Ritual, Community and New Media* (London: Routledge, 2017); Heidi A. Campbell, *Exploring Religious Community Online: We Are One in the Network* (New York: Peter Lang, 2005).
51. Heidi A. Campbell, "Who's Got the Power? Religious Authority and the Internet," *Journal of Computer-Mediated Communication* 33, no. 5 (2007); Heidi A. Campbell and Oren Golan, "Creating Digital Enclaves: Negotiation of the Internet among Bounded Religious Communities," *Media, Culture & Society* 33, no. 5 (2011), https://doi.org/10.1177/0163443711404464; Karine Barzilai-Nahon and Gad Barzilai, "Cultured Technology: The Internet and Religious Fundamentalism," *Information Society* 21, no. 1 (2005), https://doi.org/10.1080/01972240590895892; Jan Scholz et al., "Listening Communities? Some Remarks on the Construction of Religious Authority in Islamic Podcasts," *Welt des Islams* 48, nos. 3–4 (2008), https://doi.org/10.1163/157006008X364721; Laura Busch, "To Come to a Correct Understanding of Buddhism: A Case Study on Spiritualizing Technology, Religious Authority, and the Boundaries of Orthodoxy and Identity in a Buddhist Web Forum," *New Media & Society* 13, no. 1 (2011), https://doi.org/10.1177/1461444810363909; Marta Kolodziejska and A. L. P. Arat, "Religious Authority Online: Catholic Case Study in Poland," *Religion & Society in Central & Eastern Europe* 9, no. 1 (2016).
52. Although focusing on charismatic-Pentecostal churches, which are somewhat outside those discussed here, Christerson and Flory's work highlights how independent churches with loose connections are growing while churches with traditional structures are in decline. Brad Christerson and Richard Flory, *The Rise of Network Christianity: How Independent Leaders Are Changing the Religious Landscape* (Oxford: Oxford University Press, 2017).
53. Heidi A. Campbell and Mia Lövheim, "Rethinking the Online-Offline Connection in the Study of Religion Online," *Information, Communication & Society* 14, no. 8 (2011): 1092.
54. Tim Hutchings, "Studying Apps: Research Approaches to the Digital Bible," in *Digital Methodologies in the Sociology of Religion*, ed. Sariya Cheruvallil-Contractor and Suda Shakkour (New York: Bloomsbury, 2015).
55. Heidi A. Campbell, "Surveying Theoretical Approaches within Digital Religion Studies," *New Media & Society* 19, no. 1 (2016), https://doi.org/10.1177/1461444816649912;; Heidi A. Campbell, "Religious Communication and Technology," *Annals of the International Communication Association* 41, nos. 3–4 (2017), https://doi.org/10.1080/23808985.2017.1374200; Mia Lövheim and Heidi A. Campbell, "Considering Critical Methods and Theoretical Lenses in Digital Religion Studies," *New Media & Society* 19, no. 1 (2017), https://doi.org/10.1177/1461444816649911.

56. Heidi A. Campbell, *When Religion Meets New Media* (New York: Routledge, 2010), 60–62.
57. Campbell, *When Religion Meets New Media*, 114.
58. Ruth Tsuria et al., "Approaches to Digital Methods in Studies of Digital Religion," *Communication Review* 20, no. 2 (2017), https://doi.org/10.1080/10714421.2017.1304137; Lövheim and Campbell, "Considering Critical Methods."
59. The companies that produced these applications often changed the way they branded the application's name over time, including capitalization and spacing. For example, WordSEARCH, as it was called in its first release, is now Wordsearch Bible. I have attempted to use the name that was in use at the time under discussion.
60. Philanthropist and owner of Hobby Lobby and Mardel Christian and Educational Supply, Mart Green, donated a large sum of money to create the Digital Bible Library, which powers YouVersion. Bill Roberts, "Mart Green Partners with Wycliffe, Biblica, Bible Society on Digital Bible Library," *Charisma News*, December 14, 2012, http://www.charismanews.com/us/34940-mart-green-partners-with-wycliffe-biblica-bible-society-on-digital-bible-library. YouVersion has also partnered with the television series *The Bible* for an undisclosed sum. Alex Murashko, "'The Bible' Producers Announce Upcoming YouVersion Companion App," *Christian Post*, February 8, 2013, http://www.christianpost.com/news/the-bible-producers-announce-upcoming-youversion-companion-app-89696/; Tim Hutchings, "Now the Bible Is an App: Digital Media and Changing Patterns of Religious Authority," in *Religion, Media, and Social Change*, ed. Kennet Granholm, Marcus Moberg, and Sofia Sjö (New York: Routledge, 2014), 151, notes the presence of a donation page on the main website.
61. Jeremy Weber, "Why Zondervan Bought BibleGateway.com," *Christianity Today*, November 6, 2008, http://www.christianitytoday.com/ct/2008/novemberweb-only/145-41.0.html.
62. About Faithlife, retrieved May 15, 2015, from https://faithlife.com/about. Full disclosure: as I explore in subsequent chapters, my present employer, Dallas Theological Seminary (DTS), developed a software packaged called CDWord in 1989 that was sold to Logos Research Systems. I was not involved in its creation or sale, but in 2013, DTS entered a partnership with Logos to provide software and library resources to all DTS students, and I did help with that project.
63. The president of Logos Bible Software, which makes both desktop and mobile products, once remarked that Logos is for "the pastor and that one guy," referring to those interested in studying the original languages of the Bible. The company has since expanded into product lines, including mobile applications, meant to attract, in the president's words, "the other ninety-eight people in the church."
64. Michael O. Emerson and Karen Chai Kim, "Multiracial Congregations: An Analysis of Their Development and a Typology," *Journal for the Scientific Study of Religion* 42, no. 2 (2003). On the neighborhood demographics see Jed Kolko, "America's Most Diverse Neighborhoods and Metros," *Forbes*, November 13, 2012, http://www.forbes.com/sites/trulia/2012/11/13/finding-diversity-in-america/.
65. Eric Reed, "The Best of Today's Preachers," *Christianity Today Pastors*, February 6, 2002, https://www.christianitytoday.com/pastors/2002/february-online-only/cln20

206.html; LifeWay Research, "Protestant Pastors Name Graham Most Influential Living Preacher," *LifeWay Research*, February 2, 2010, https://lifewayresearch.com/2010/02/02/protestant-pastors-name-graham-most-influential-living-preacher/.

66. Jeff Brumley, "Name Changes Challenge Churches on Baptist Identity," *Baptist Global News*, October 7, 2014, https://baptistnews.com/ministry/congregations/item/29312-name-changes-challenge-churches-on-baptist-identity; Susan Montoya, "Churches Take 'Baptist' from Name," *AP News Archive Beta*, January 12, 1999, http://www.apnewsarchive.com/1999/Churches-Take-Baptist-From-Name/id-841aabf08eedd8c220aa7683cfa72f37.
67. Peter H. Davids, *The Letters of 2 Peter and Jude* (Grand Rapids, MI: Eerdmans, 2006), 7.
68. Thomas R. Schreiner, *1, 2 Peter, Jude* (Nashville: Broadman & Holman Publishers, 2003), 403.
69. D. J. Rowston, "The Most Neglected Book in the New Testament," *New Testament Studies* 21, no. 4 (1975): 554.

Chapter 3

1. "What is an evangelical? That question brings more people to the website the of the National Association of Evangelicals than another other search." National Association of Evangelicals, "Context," *Evangelicals*, Winter 2017, 5.
2. David W. Bebbington, *Evangelicalism in Modern Britain: A History from the 1730s to the 1980s* (London: Unwin Hyman, 1989), 1.
3. Samuel Crossley, "Recent Developments in the Definition of Evangelicalism," *Foundations (Affinity)* 70 (2016): 112.
4. Bijker, Hughes, and Pinch, "Introduction," 6–8.
5. Other relevant groups include nonevangelical Bible readers, along with the broader categories of hardware creators, software developers, and general users of digital products. These groups will be considered tangentially in the following chapters, but the focus will remain on evangelical developers and users as defined in chapter 2.
6. Heidi A. Campbell and Kyong James Cho, "Religious Use of Mobile Phones," in *Encyclopedia of Mobile Phone Behavior*, ed. Zheng Yan (Hersey, PA: Information Science Reference, 2015).
7. Campbell, *When Religion Meets New Media*, 50.
8. Kenneth J. Stewart, "Did Evangelicalism Predate the Eighteenth Century? An Examination of David Bebbington's Thesis," *Evangelical Quarterly* 77, no. 2 (2005): 152.
9. Timothy Weber, "Premillenialism and the Branches of Evangelicalism," in *The Variety of American Evangelicalism*, ed. Donald W. Dayton and Robert K. Johnston (Nashville: University of Tennessee Press, 2001), 12.
10. Crossley, "Recent Developments," 112.

11. Thomas More, *The Yale Edition of the Complete Works of St. Thomas More*, vol. 8, ed. Louis A. Schuster, Richard C. Marius, and James P. Lusardi (New Haven: Yale University Press, 1973).
12. Linford D. Fisher, "Evangelicals and Unevangelicals: The Contested History of a Word, 1500–1950," *Religion and American Culture* 26, no. 2 (Summer 2016): 187, 88, https://doi.org/10.1525/rac.2016.26.2.184.
13. Bebbington, *Evangelicalism in Modern Britain*, 20–3.
14. Norman Etherington, ed., *Missions and Empire* (Oxford: Oxford University Press, 2005), writes about the connection between bringing the gospel and bringing Western technology and colonialism.
15. Crossley, "Recent Developments," 115.
16. Crawford Gribben, "Puritan Subjectivities: The Converstion Debate in Cromwellian Dublin," in *Converts and Conversion in Ireland, 1650–1850*, ed. Michael L. Brown, McGrath Carlies Ivar, and Tom P. Power (Dublin: Four Courts Press, 2005), 86.
17. Stewart challenges the Bebbington thesis, arguing that the four elements of the quadrilateral were, in fact, associated with one another prior to the 1730s and that evangelicalism was not a response to unique factors the eighteenth century, but rather, "arose in light of recurring perennial factors." Stewart, "Did Evangelicalism Predate," 138, 42.
18. Molly Worthen, *Apostles of Reason: The Crisis of Authority in American Evangelicalism* (Oxford: Oxford University Press, 2016), 6.
19. Mark A. Noll, "What Is 'Evangelical'?," in *The Oxford Handbook of Evangelical Theology*, ed. Gerald R. McDermott (Oxford: Oxford University Press, 2010), 24.
20. W. Reginald Ward, *Early Evangelicalism: A Global Intellectual History, 1670–1789* (Cambridge: Cambridge University Press, 2006).
21. Catherine A. Brekus, *Sarah Osborn's World: The Rise of Evangelical Christianity in Early America* (New Haven: Yale University Press, 2013), 7.
22. Kristin Kobes Du Mez, *Jesus and John Wayne: How White Evangelicals Corrupted a Faith and Fractured a Nation* (New York: Liveright Publishing Corporation, 2020).
23. Alister McGrath, *Evangelicalism and the Future of Christianity* (London: Hodder & Stoughton, 1993), 27.
24. George Marsden offers the following tongue-in-cheek distinction between fundamentalism and evangelicalism: "My own unscientific shorthand for this broader usage is that a fundamentalist (or a fundamentalistic evangelical) is 'an evangelical who is angry about something.'" George M. Marsden, *Fundamentalism and American Culture*, 2nd ed. (Oxford: Oxford University Press, 2006), 235.
25. David Allan Hubbard, *What We Evangelicals Believe* (Pasadena, CA: Fuller Seminary Press, 1991), 9.
26. Crossley, "Recent Developments," 112.
27. George M. Marsden, *Reforming Fundamentalism: Fuller Seminary and the New Evangelicalism* (Grand Rapids, MI: Eerdmans, 1987), 230.
28. Meacham, "The Editor's Desk."
29. "That Old Time Religion: The Evangelical Empire," *Time*, December 26, 1977, http://content.time.com/time/subscriber/article/0,33009,919227-2,00.html.

30. These biases included understanding evangelicalism primarily as an expression of social class and being unworthy of study because it did not cohere with the liberal outlook of sociology. R. Stephen Warner, "Theoretical Barriers to the Understanding of Evangelical Christianity," *Sociological Analysis* 40, no. 1 (1977).
31. Warner, "Theoretical Barriers," 2.
32. Steve Bruce, "Identifying Conservative Protestantism," *Sociological Analysis* 44, no. 1 (1983).
33. James Davison Hunter, "Operationalizing Evangelicalism: A Review, Critique & Proposal," *Sociological Analysis* 42, no. 4 (1981).
34. Nancy Ammerman, "Operationalizing Evangelicalism: An Amendment," *Sociological Analysis* 43, no. 2 (1982): 170.
35. Matthew Avery Sutton, *Jerry Falwell and the Rise of the Religious Right: A Brief History with Documents* (Boston: Bedford / St. Martin's, 2013); Gabriel Fackre, *The Religious Right and Christian Faith* (Grand Rapids, MI: Eerdmans, 1982); Gary E. McCuen, *The Religious Right* (Hudson, WI: G.E. McCuen Publications, 1989); Clyde Wilcox and Carin Robinson, *Onward Christian Soldiers? The Religious Right in American Politics*, 4th ed. (Boulder, CO: Westview Press, 2011), 6.
36. Stephen D. Johnson and Joseph B. Tamney, "The Christian Right and the 1980 Presidential Election," *Journal for the Scientific Study of Religion* 21, no. 2 (1982), https://doi.org/10.2307/1385498; Walter H. Capps, *The New Religious Right: Piety, Patriotism, and Politics* (Columbia: University of South Carolina Press, 1990).
37. Ed Stetzer, "Defining Evangelicals in Research," *The Evangelicals* 3, no. 3 (2017–18): 12.
38. Noll, "What Is Evangelical," 19.
39. "Religious Landscape Study," 2014, accessed March 17, 2017, http://www.pewforum.org/religious-landscape-study/.
40. Daniel Cox and Robert P. Jones, *America's Changing Religious Identity*, Public Religion Research Institute, September 6, 2017, https://www.prri.org/research/american-religious-landscape-christian-religiously-unaffiliated/.
41. "How the Faithful Voted: A Preliminary 2016 Analysis," Pew Research, updated November 9, 2016, http://www.pewresearch.org/fact-tank/2016/11/09/how-the-faithful-voted-a-preliminary-2016-analysis/.
42. Mark Labberton, "Opinion: Are Evangelicals Today More Devoted to Trump and the Republicans Than the Gospel?," *LA Times*, August 26 2016, http://www.latimes.com/opinion/readersreact/la-ol-le-trump-evangelicals-white-supremacy-20170826-story.html.
43. For example, Labberton and other evangelicals contributed to a book discussing the ongoing value of the term and the movement. Mark Labberton, *Still Evangelical? Insiders Reconsider Political, Social and Theological Meaning* (Downers Grove, IL: InterVarsity Press, 2018).
44. Trip Gabriel, "Donald Trump, Despite Impieties, Wins Hearts of Evangelical Voters," *New York Times*, February 28, 2016, https://www.nytimes.com/2016/02/28/us/politics/donald-trump-despite-impieties-wins-hearts-of-evangelical-voters.html; Myriam Renaud, "Myths Debunked: Why Did White Evangelical Christians Vote for Trump?," University of Chicago Divinity School, Martin Marty Center for the

Public Understanding of Religion, January 19, 2017, https://divinity.uchicago.edu/sightings/myths-debunked-why-did-white-evangelical-christians-vote-trump.

45. Christian Smith and Michael O. Emerson, *American Evangelicalism: Embattled and Thriving* (Chicago: University of Chicago Press, 1998), 218.
46. Doreen M. Rosman, *Evangelicals and Culture*, 2nd ed. (Cambridge: James Clarke, 2012).
47. Marsden followed Fuller Theological Seminary's change of position regarding inerrancy. Marsden, *Reforming Fundamentalism*, 111–13, 213–14, 246. On open theism, see David Hillborn, "Principled Unity or Pragmatic Compromise? The Challenge of Pan-Evangelical Theology," *Evangel* 22, no. 3 (2004): 81–82. As an example of evangelical reaction to marriage, see the Council for Biblical Manhood and Womanhood's "Nashville Statement" (nashvillestatement.com), released in 2017, accessed May 18, 2022.
48. Albert Mohler, "Confessional Evangelicalism," in *Four Views on the Spectrum of Evangelicalism*, ed. Collin Hansen (Grand Rapids, MI: Zondervan, 2011), 73–75.
49. J. C. Ryle, *Evangelical Religion: What It Is, and What It Is Not* (London: William Hunt, 1867), 7–10.
50. Harold Ockenga and Leslie R. Marston, *Our Evangelical Faith* (Whitefish, MT: Literary Licensing, 1946).
51. His other elements include the majesty of Jesus Christ; the lordship of the Holy Spirit; the necessity of conversion; the priority of evangelism; the importance of fellowship. J. I. Packer, *The Evangelical Anglican Identity Problem: An Analysis* (Latimer, England: Latimer House, 1978), 15–23.
52. John R. W. Stott, *Evangelical Truth: A Personal Plea for Unity, Integrity and Faithfulness* (Carlisle, PA: Langham Global Library, 2013), 28, 75.
53. J. I. Packer and Thomas C. Oden, *One Faith: The Evangelical Consensus* (Westmont, IL: IVP Books, 2004), 19.
54. Brian Malley, *How the Bible Works: An Anthropological Study of Evangelical Biblicism* (Walnut Creek, CA: AltaMira Press, 2004), 35.
55. Robert Wuthnow, *Sharing the Journey: Support Groups and America's New Quest for Community* (London: Free Press, 1996).
56. Bielo, *Words upon the Word*; James S. Bielo, "On the Failure of 'Meaning': Bible Reading in the Anthropology of Christianity," *Culture and Religion* 9, no. 1 (2008).
57. Nancy Tatom Ammerman, *Congregation & Community* (New Brunswick, NJ: Rutgers University Press, 1997).
58. Mathew Guest, *Evangelical Identity and Contemporary Culture: A Congregational Study in Innovation* (Milton Keynes: Paternoster, 2007), 190–95.
59. "Blacks More Likely Than Others in U.S. to Read the Bible Regularly, See It as God's Word," Pew Research, updated May 7, 2018, http://www.pewresearch.org/fact-tank/2018/05/07/blacks-more-likely-than-others-in-u-s-to-read-the-bible-regularly-see-it-as-gods-word/.
60. Pew also notes that Jehovah's Witnesses (88%) and Mormons (77%) read scripture weekly at higher rates than evangelicals (63%), and also have stronger beliefs about scripture as God's word and attend weekly studies more often (Pew Research, "Blacks

More Likely"). Some evangelicals have sought to find more common ground with Mormons, although they also desire to remain distinctive. Richard J. Mouw, *Talking with Mormons: An Invitation to Evangelicals* (Grand Rapids, MI: Eerdmans, 2012).

61. Buck Gee and Denise Peck, *The Illusion of Asian Success: Scant Progress for Minorities in Cracking the Glass Ceiling from 2007–2015* (Ascend Foundation Research, 2017), accessed May 18, 2022, https://c.ymcdn.com/sites/www.ascendleadership.org/resource/resmgr/research/TheIllusionofAsianSuccess.pdf.
62. Tanya Luhrmann, *When God Talks Back: Understanding the American Evangelical Relationship with God* (New York: Vintage, 2012), 131.
63. Rosman, *Evangelicals and Culture*, 4.
64. Richard G. Kyle, *Evangelicalism: An Americanized Christianity* (New Brunswick, NJ: Transaction Publishers, 2006), 167.
65. Richard G. Kyle, *Popular Evangelicalism in American Culture* (London: Routledge, 2017).
66. Corrina Laughlin, "'What God Gave to Us': Digital Habitus and the Shifting Social Imaginary of American Evangelicalism" (PhD diss., University of Pennsylvania, 2018), 24–25.
67. Ian G. Barbour, *Ethics in an Age of Technology*, vol. 2 (San Francisco: HarperCollins, 1993).
68. John Ferre, "The Media of Popular Piety," in *Mediating Religion: Conversations in Media, Religion and Culture*, ed. J. Mitchell and S. Marriage (London: T&T Clark, 2003), 83–92.
69. Billy Graham, *Just as I Am: The Autobiography of Billy Graham*, 10th anniversary ed. (New York: HarperOne, 2007), 432.
70. Adam McLane, "The Opportunities Smartphones Present for Your Church," vol. 4, no. 1 (Spring/Summer), *The Evangelicals*, 2018, 15.
71. Josh Hall, "Technophobic Pastors of Tech-Addicted Churches," vol. 4, no. 1 (Spring/Summer), *The Evangelicals*, 2018, 19.
72. Wade Clark Roof, *Spiritual Marketplace: Baby Boomers and the Remaking of American Religion* (Princeton, NJ: Princeton University Press, 1999), 25.
73. Randall Herbert Balmer, "The Wireless Gospel: Sixty-Two Years Ago, Back to the Bible Joined the Radio Revolution; Now It Is Finding New Media for Its Old Message. A Case Study in Evangelicals' Love Affair with Communications Technology," *Christianity Today* 45, no. 3 (2001): 48.
74. Heather Gonzalez, "NAE's Beginning," vol. 3, no. 3 (Winter), *Evangelicals*, 2017, 20.
75. Kenneth W. M. Wozniak, "Evangelicals and the Ethics Of Information technology," *Journal of the Evangelical Theological Society* 28, no. 3 (1985): 336.
76. Stephen J. Monsma, *Responsible Technology: A Christian Perspective* (Grand Rapids, MI: Eerdmans, 1986), 3, 24.
77. Quentin J. Schultze, *Habits of the High-Tech Heart: Living Virtuously in the Information Age* (Grand Rapids, MI: Baker Books, 2002); Shane Hipps, *Flickering Pixels: How Technology Shapes Your Faith* (Grand Rapids, MI: Zondervan, 2009); Shane Hipps, *The Hidden Power of Electronic Culture: How Media Shapes Faith, the Gospel, and Church* (El Cajon, CA: Youth Specialties, 2006); T. David Gordon,

Why Johnny Can't Sing Hymns: How Pop Culture Rewrote the Hymnal (Phillipsburg, NJ: P & R Pub., 2010); T. David Gordon, *Why Johnny Can't Preach: The Media Have Shaped the Messengers* (Phillipsburg, NJ: P & R Pub., 2009).

78. Geoff Surratt, Greg Ligon, and Warren Bird, *The Multi-site Church Revolution: Being One Church—in Many Locations* (Grand Rapids, MI: Zondervan, 2006).
79. Douglas Estes, *SimChurch: Being the Church in the Virtual World* (Grand Rapids, MI: Zondervan, 2009).
80. Tim Challies, *The Next Story: Life and Faith after the Digital Explosion* (Grand Rapids, MI: Zondervan, 2011), 76.
81. Derek C. Schuurman, *Shaping a Digital World: Faith, Culture and Computer Technology* (Downers Grove, IL: IVP Academic, 2013); Craig M. Gay, *Modern Technology and the Human Future: A Christian Appraisal* (Downers Grove, IL: InterVarsity Press, 2018); Andrew Byers, *TheoMedia: The Media of God and the Digital Age* (Havertown: Lutterworth Press, 2014).
82. *Ecclesiology for a Digital Church*, ed. John Dyer and Heidi A. Campbell (London: SCM Press, 2022).
83. Calvin Miller, *The Vanishing Evangelical: Saving the Church from Its Own Success by Restoring What Really Matters* (Grand Rapids, MI: Baker Books, 2013).
84. "Church and Technology: A Survey of Ontario Churches," *Tyndale: The Magazine*, Fall 2011.
85. Brekus, *Sarah Osborn's World*, 10.
86. "Four-in-Ten Americans Credit Technology with Improving Life Most in the Past 50 Years," Pew Research Center, 2017, accessed November 24, 2018, http://www.pewresearch.org/fact-tank/2017/10/12/four-in-ten-americans-credit-technology-with-improving-life-most-in-the-past-50-years/.
87. Lindsay Nicolet, "The Bible as an App," vol. 4, no. 1, *Light*, 2018, 17.
88. Juzwik Mary M. Juzwik, "American Evangelical Biblicism as Literate Practice: A Critical Review," *Reading Research Quarterly* 49, no. 3 (2014), https://doi.org/10.1002/rrq.72, argues that evangelicals' use of the Bible is a kind of practice of its own.
89. "Evangelism Never Changes, but the Methods Do," updated February 6, 2015, https://www.christianitytoday.com/edstetzer/2015/february/evangelism-never-changes-but-never-stays-same.html.
90. Tecoy M. Porter Sr., *Faith To Innovate: 21st Century Tools & Strategies for Leadership Transformation* (Elk Grove, CA: Inner Treasure Press, 2015), 60.
91. Richard W. Flory and Donald E. Miller, *Finding Faith: The Spiritual Quest of the Post-boomer Generation* (New Brunswick, NJ: Rutgers University Press, 2008); Dann Wigner, *A Sociology of Mystic Practices: Use and Adaptation in the Emergent Church* (Eugene, OR: Pickwick Publications, 2018).
92. Younger evangelicals today appear to be less interested in issues of like sexuality and abortion and are more likely to be vocal about issues like immigration and climate change. Eliza Griswold, "Millennial Evangelicals Diverge from Their Parents' Beliefs," *New Yorker*, August 27, 2018, https://www.newyorker.com/news/on-religion/millennial-evangelicals-diverge-from-their-parents-beliefs.

93. Though it can be seen in other cultures such as Korea and Guatemala. Jean-Paul Baldacchino, "Markets of Piety and Pious Markets: The Protestant Ethic and the Spirit of Korean Capitalism," *Social Compass* 59, no. 3 (2012), https://doi.org/10.1177/0037768612449721; Peter L. Berger, "Max Weber Is Alive and Well, and Living in Guatemala: The Protestant Ethic Today," *Review of Faith & International Affairs* 8, no. 4 (2010).
94. Kathryn Tanner, *Christianity and the New Spirit of Capitalism* (New Haven: Yale University Press, 2019); Randall Herbert Balmer, *Blessed Assurance: A History of Evangelicalism in America* (Boston: Beacon Press, 1999), 11.
95. Wyndy Corbin, "The Impact of the American Dream on Evangelical Ethics," *Cross Currents* 55, no. 3 (Fall 2005): 346.
96. Wayne demonstrated that there remains a strong overlap between elements of the Protestant work ethic, such as individualism, asceticism, and industriousness, and new emerging values. F. Stanford Wayne, "An Instrument to Measure Adherence to the Protestant Ethic and Contemporary Work Values," *Journal of Business Ethics* 8 (1989); Robert W. Green, *Protestantism and Capitalism: The Weber Thesis and Its Critics* (Boston: Heath, 1959).
97. James S. Bielo, "'The Mind of Christ': Financial Success, Born-Again Personhood, and the Anthropology of Christianity," *Ethnos: Journal of Anthropology* 72, no. 3 (2007): 326, https://doi.org/10.1080/00141840701576935.
98. Bethany Moreton, *To Serve God and Wal-Mart: The Making of Christian Free Enterprise* (Cambridge, MA: Harvard University Press, 2010).
99. William E. Connolly, *Capitalism and Christianity, American Style* (Durham, NC: Duke University Press Books, 2008), 48. Connolly's initial examples are tied to the era of Left Behind and the Bush-Cheney administration, but the concept continues in the era of Jerry Falwell Jr. and the Trump presidency. William E. Connolly, "The Evangelical-Capitalist Resonance Machine," *Political Theory* 33, no. 6 (2005), http://www.jstor.org/stable/30038467.
100. Jon Bialecki, "Between Stewardship and Sacrifice: Agency and Economy in a Southern California Charismatic Church," *Entre gestion et sacrifice* 14, no. 2 (2008), https://doi.org/10.1111/j.1467-9655.2008.00507.x.
101. Kevin D. Dougherty et al., "A Religious Profile of American Entrepreneurs," *Journal for the Scientific Study of Religion* 52, no. 2 (2013), https://doi.org/10.1111/jssr.12026.
102. Kevin M. Kruse, *One Nation under God: How Corporate America Invented Christian America* (New York: Basic Books, 2015), http://www.h-net.org/reviews/showrev.php?id=44131.
103. D. Michael Lindsay, *Faith in the Halls of Power: How Evangelicals Joined the American Elite* (Oxford: Oxford University Press, 2007).
104. Ayantunji Gbadamosi, "Exploring the Growing Link of Ethnic Entrepreneurship, Markets, and Pentecostalism in London (UK): An Empirical Study," *Society and Business Review* 10 (2015).
105. Nanlai Cao, "Boss Christians: The Business of Religion in the 'Wenzhou Model' of Christian Revival," *China Journal* 59 (2008), https://doi.org/10.2307/20066380.
106. Smith argues that secularization in the United States at the turn of the twentieth century was driven not by modernity itself, but by elite intellectual class.

Christian Smith, *The Secular Revolution: Power, Interests, and Conflict in the Secularization of American Public Life* (Berkeley: University of California Press, 2003); Pippa Norris and Ronald Inglehart, *Sacred and Secular: Religion and Politics Worldwide* (Cambridge: Cambridge University Press, 2011), 3–82; Callum G. Brown, *The Death of Christian Britain: Understanding Secularisation, 1800–2000* (London: Routledge, 2001).

107. This line of reasoning draws on rational choice theory as an alternative to secularization. Grace Davie, *The Sociology of Religion*, 2nd ed. (London: Sage, 2013), 67–89. However, other sociologists are critical of rational choice theory for not being viable outside Christianity. Stephen Sharot, "Beyond Christianity: A Critique of the Rational Choice Theory of Religion from a Weberian and Comparative Religions Perspective," *Sociology of Religion* 63, no. 4 (2002), https://doi.org/10.2307/3712301.
108. The often-quoted version of this is, "[People] like to become Christians without crossing racial, linguistic or class barriers." Donald A. McGavran, *Understanding Church Growth* (Grand Rapids, MI: Eerdmans, 1970), 198.
109. Bruce W. Fong, *Racial Equality in the Church: A Critique of the Homogeneous Unit Principle in Light of a Practical Theology Perspective* (Lanham, MD: University Press of America, 1996); Mark DeYmaz, *Building a Healthy Multi-ethnic Church: Mandate, Commitments, and Practices of a Diverse Congregation* (San Francisco: Jossey-Bass / John Wiley, 2007).
110. Patrick focuses on the importance of character: "The most critical human component of every church plant is the planter." Darrin Patrick, *Church Planter: The Man, the Message, the Mission* (Wheaton, IL: Crossway, 2010)..
111. In addition to these networks, many successful church planters have written books about how to replicate their success. Nelson Searcy and Kerrick Thomas, *Launch: Starting a New Church from Scratch*, rev. ed. (Grand Rapids, MI: Baker Books, 2017); Mike McKinley, *Church Planting Is for Wimps: How God Uses Messed-Up People to Plant Ordinary Churches That Do Extraordinary Things* (Wheaton, IL: Crossway Books, 2010); Jervis David Payne, *Apostolic Church Planting: Birthing New Churches from New Believers* (Downers Grove, IL: InterVarsity Press, 2015). These also include books designed to help the family of the church planter. Christine Hoover, *The Church Planting Wife: Help and Hope for Her Heart* (Chicago: Moody Publishers, 2013).
112. "Church Planting 1: The Movement," Startup from Gimlet Media, updated July 6, 2018, https://www.gimletmedia.com/startup/church-planting-1-the-movement.
113. Charles Brown, "Selling Faith: Marketing Christian Popular Culture to Christian and Non-Christian Audiences," *Journal of Religion and Popular Culture* 24, no. 1 (Spring 2012), https://doi.org/10.3138/jrpc.24.1.113.
114. Sean Benesh, *Intrepid: Navigating the Intersection of Church Planting + Social Entrepreneurship* (n.p.: Missional Challenge, 2018).
115. Andy Crouch, *Culture Making: Recovering Our Creative Calling* (Downers Grove, IL: IVP Books, 2013), 68–72.
116. John MacArthur, *Ashamed of the Gospel: When the Church Becomes Like the World*, 3rd ed. (Wheaton, IL: Crossway Books, 2010), 27.

117. Nelson's critique is not centered on adopting cultural goods like technology, but rather on adopting attitudes about goals in life and the place of faith in achieving those goals. Peter K. Nelson, "Impractical Christianity: Faith Really Begins to Make a Difference When It Stops 'Working,'" *Christianity Today* 49, no. 9 (2005): 80.
118. Samuel R. Schutz, "The Truncated Gospel in Modern Evangelicalism: A Critique and Beginning Reconstruction," *Evangelical Review of Theology* 33, no. 4 (2009): 292.
119. By "decision," Shultz means that a person who hears or reads a simplified gospel presentation will pray a make a "decision" to become a Christian, which happens when he or she prays a "sinner's prayer" that results in an evangelical "conversion."
120. Allan Tate, *Essays of Four Decades* (Wilmington: ISI Books, 1999), 6.
121. Tate also felt that the radio and other communication technology negatively affected human love. "Communication that is not also communion is incomplete. We *use* communication; we *participate* in communion." Tate, *Essays of Four Decades*, 10.
122. A. J. Conyers, "Three Sources of the Secular Mind," *Journal of the Evangelical Theological Society* (June 1, 1998): 313.
123. Mark A. Noll, "What Has Wheaton to Do with Jerusalem? Lessons from Evangelicals for the Reformed," *Reformed Journal* 32, no. 5 (1982): 9.
124. Noll, "Wheaton," 10.
125. John Mark Hicks, "Numerical Growth in the Theology of Acts: The Role of Pragmatism, Reason and Rhetoric," paper presented at the Forty-Seventh National Conference of the Evangelical Theological Society, Philadelphia, PA, November 16–18, 1995\.
126. Hillborn, "Principled Unity or Pragmatic Compromise," 80.
127. E. Glenn Hinson, "William Carey and Ecumenical Pragmatism," *Journal of Ecumenical Studies* 17, no. 2 (1980): 73.
128. Hinson, "William Carey," 82.
129. Stephen Tomkins, *John Wesley: A Biography* (Grand Rapids, MI: Eerdmans, 2003), 160.
130. William C. Guerrant, *Organic Wesley: A Christian Perspective on Food, Farming, and Faith* (Franklin, TN: Seedbed Publishing, 2015).
131. Melvin Eugene Page and Penny M. Sonnenburg, *Colonialism: An International, Social, Cultural, and Political Encyclopedia*, 3 vols. (Santa Barbara, CA: ABC-CLIO, 2003).
132. Robert Wuthnow, *Boundless Faith: The Global Outreach of American Churches* (Berkeley: University of California Press, 2009), 2.
133. As mentioned above, there are many angles of critique on the missions movement, especially regarding its connection to colonialism. More recently, Haynes argues that short-term mission trips are often setup for the benefit of the those who are going rather than those who are being visited. Robert Ellis Haynes, *Consuming Mission: Towards a Theology of Short-Term Mission and Pilgrimage* (Eugene, OR: Pickwick Publications, 2018).
134. Gary Scott Smith, *Faith and the Presidency: from George Washington to George W. Bush* (Oxford: Oxford University Press, 2006), 293–324.

135. "Clinton Maintains Double-Digit (51% vs. 36%) Lead over Trump," Public Religion Research Institute / Brookings Survey, updated October 19, 2016, https://www.prri.org/research/prri-brookings-oct-19-poll-politics-election-clinton-double-digit-lead-trump/.
136. Thomas B. Edsall, "Trump Says Jump: His Supporters Ask, How High?," *New York Times*, September 14, 2017, https://www.nytimes.com/2017/09/14/opinion/trump-republicans.html.
137. Jessica Martínez and Gregory A. Smith, "How the Faithful Voted: A Preliminary 2016 Analysis," Pew Research, updated November 9, 2016, http://www.pewresearch.org/fact-tank/2016/11/09/how-the-faithful-voted-a-preliminary-2016-analysis/; Gabriel, "Donald Trump, Despite Impieties."
138. Researchers have shown that many churches perpetuate racial segregation by "executing what we term 'race tests,' on incoming people of color." Glenn E. Bracey II and Wendy Leo Moore, "Race Tests: Racial Boundary Maintenance in White Evangelical Churches," *Sociological Inquiry* 87, no. 2 (May 2017): 282, http://onlinelibrary.wiley.com/wol1/doi/10.1111/soin.12174/full. The "ChurchToo" movement arose in response to sexual abuse within the Christian churches. Lea Karen Kivi, "#CHURCHTOO," *America* 219, no. 12 (November 26, 2018). During this time there were several evangelical sex abuse scandals including former youth pastor Andy Savage. Kate Shellnut, "#ChurchToo: Andy Savage Resigns from Megachurch over Past Abuse," *Christianity Today*, March 18, 2018, https://www.christianitytoday.com/news/2018/march/andy-savage-resigns-abuse-megachurch-standing-ovation.html. Evangelical opinion has ranged from supporting gay rights in the 1980s to continued antihomosexual rhetoric and practices after the United States recognized gay marriage. Nancy Hardesty, "Evangelical Women Face Their Homophobia," *Christian Century* 103, no. 26 (1986); Jeremy N. Thomas and Andrew L. Whitehead, "Evangelical Elites' Anti-homosexuality Narratives as a Resistance Strategy against Attribution Effects," *Journal for the Scientific Study of Religion* 54, no. 2 (2015), https://doi.org/10.1111/jssr.12188. Stetzer urges evangelicals not to fall for Trump's rhetoric about immigration. Ed Stetzer, "Fellow Evangelicals: Stop Falling for Trump's Anti-immigrant Rhetoric," *Vox*, November 2, 2018, https://www.vox.com/policy-and-politics/2018/11/6/18066116/trump-caravan-evangelical-voters. John MacArthur's university has been accused by accreditors of abuses of power. Eric Kelderman and Dan Bauman, "'Fear, Intimidation, Bullying': Inside One of the Most Scathing Accreditation Reports in Recent Memory," *Chronicle*, December 21, 2018. Perry argues that evangelicals use pornography at only slightly lower rates than the rest of the general public, but that this causes them greater distress because of their moral commitments. Samuel L. Perry, *Addicted to Lust: Pornography in the Lives of Conservative Protestants* (New York: Oxford University Press, 2019).
139. Notable evangelical institutions that have spoken and written against Donald Trump include *Christianity Today* and the Gospel Coalition and individuals like Albert Mohler of the Southern Baptist Theological Seminary and Russell Moore of the Ethics and Religious Liberty Commission. "The Non-Trump evangelicals," *The Economist*, April 19, 2018, https://www.economist.com/united-states/2018/04/19/the-non-trump-evangelicals; Sarah Pulliam Bailey, "Dozens of Evangelical Leaders

Meet to Discuss How Trump Era Has Unleashed 'Grotesque Caricature' of Their Faith," *Washington Post*, April 16 2018, https://www.washingtonpost.com/news/acts-of-faith/wp/2018/04/12/when-you-google-evangelicals-you-get-trump-high-profile-evangelicals-will-meet-privately-to-discuss-their-future/.

140. Jerry Falwell Jr., "Moderate Republicans Make My Blood Boil," April 29, 2017, http://video.foxnews.com/v/5416587832001/.
141. "Honored to introduce @realDonaldTrump at religious leader summit in NYC today! He did incredible job! @beckifalwell," tweet, June 21, 2016, https://twitter.com/JerryFalwellJr/status/745325187776811008; Joe Helm, "Jerry Falwell Jr. Can't Imagine Trump 'Doing Anything That's Not Good for the Country,'" *Washington Post*, January 1, 2019, https://www.washingtonpost.com/lifestyle/magazine/jerry-falwell-jr-cant-imagine-trump-doing-anything-thats-not-good-for-the-country/2018/12/21/6affc4c4-f19e-11e8-80d0-f7e1948d55f4_story.html.
142. These include Billy Graham's son Franklin Graham, James Dobson, founder of Focus on the Family, Ronnie Floyd, former president of the Southern Baptist Convention, Ralph Reed, founder of the Faith and Freedom Coalition, and author and commentator Eric Metaxas. Kate Shellnut and Sarah Eekhoff Zylstra, "Who's Who of Trump's 'Tremendous' Faith Advisers," *Christianity Today*, June 22, 2016, https://www.christianitytoday.com/ct/2016/june-web-only/whos-who-of-trumps-tremendous-faith-advisors.html.
143. Jonathan Dudley, *Broken Words: The Abuse of Science and Faith in American Politics* (New York: Crown Publishers, 2011), 46.
144. Nina Burleigh, "Does God Believe in Trump? White Evangelicals Are Sticking with Their 'Prince of Lies,'" *Newsweek*, October 5, 2017, http://www.newsweek.com/2017/10/13/donald-trump-white-evangelicals-support-god-677587.html.
145. "The Real Origins of the Religious Right," *Politico*, updated May 27, 2014, https://www.politico.com/magazine/story/2014/05/religious-right-real-origins-107133?paginate=false.
146. Capps, *The New Religious Right*, 93.
147. David Brody and Scott Lamb, *The Faith of Donald J. Trump: A Spiritual Biography* (New York: Broadside Books, 2018).
148. Hugh Hewitt, "Why Christians Will Stick with Trump," *Washington Post*, October 5 2017, https://www.washingtonpost.com/opinions/why-christians-will-stick-with-trump/2017/10/05/7d7d2bb6-a922-11e7-850e-2bdd1236be5d_story.html.
149. Gabriel, "Donald Trump, Despite Impieties."
150. robertjeffress, August 18, 2017, tweet, https://twitter.com/robertjeffress/status/898733278215254016.
151. Robert Pear, "Trump Administration Set to Roll Back Birth Control Mandate," *New York Times*, October 5 2017, https://www.nytimes.com/2017/10/05/us/politics/trump-birth-control.html.
152. Renaud, *Myths Debunked.*
153. "Exposing America's Biggest Hypocrites: Evangelical Christians," *Huffington Post*, updated November 24, 2017, https://www.huffingtonpost.com/entry/exposing-americas-biggest-hypocrites-evangelical_us_5a184f0ee4b068a3ca6df7ad.

154. I am indebted to Mathew Guest for pointing out the evolution of pragmatism toward instrumentalism and the way in which Trump himself models this instrumentalism by appropriating evangelical language, deliberately lying, and presenting things in a way that serves his own self-interests.
155. Noll, "What Is Evangelical," 29–30.
156. Mark A. Noll, *American Evangelical Christianity: An Introduction* (Oxford: Blackwell Publishers, 2001), 2.

Chapter 4

1. These are (1) first wave: American independence to the 1820s, using early print presses from England, (2) second wave: 1820s to 1870s with advances in stereotyping, (3) third wave: 1870s to 1980s, which brought many new translations, (4) fourth wave: 1980s and beyond when Bible publishing became digitized. Paul Gutjahr, "Protestant English-Langauge Bible Publishing and Translation," in *The Oxford Handbook of the Bible in America*, ed. Paul Gutjahr (Oxford: Oxford University Press, 2018).
2. Jack Burch, "The Use of a Computer in New Testament Text Criticism," *Restoration Quarterly* 8, no. 2 (1965).
3. "According to Mark IV," *Time*, August 9, 1954, http://content.time.com/time/magazine/article/0,9171,936302,00.html.
4. Harbin, "Fiat Lux," 107.
5. Delores M. Burton, "Automated Concordances and Word Indexes: The Fifties," *Computers and the Humanities* 1, no. 15 (1981): 6.
6. Meredith Hindley, "The Rise of the Machines: NEH and the Digital Humanities: The Early Years," *Humanities* 34, no. 4 (2013), http://www.neh.gov/humanities/2013/julyaugust/feature/the-rise-the-machines.
7. "Bible Labor of Years Is Done in 400 Hours," *Life*, February 18, 1957.
8. Matthew Brook O'Donnell, "Linguistic Fingerprints or Style by Numbers? The Use of Statistics in the Discussion of Authorship of New Testament Documents," in *Linguistics and the New Testament: Critical Junctures*, ed. Stanley E. Porter and D. A. Carson (Sheffield: Sheffield Academic Press, 1999).
9. Morton argued that his computationally derived linguistic analyses that looked at the frequency of words like *kai* ("and") and *de* ("but") showed that only Romans, Galatians, Philemon, and the Corinthian letters bore a distinct enough authorial style to be attributed to a single author.
10. John Ellison, "Review of Christianity in the Computer Age," *Journal of Biblical Literature* 84, no. 4 (1965); C. Dinwoodie, "Christianity in the Computer Age," *Scottish Journal of Theology* 18, no. 2 (1965); Robert Goodwin, "Christianity in the Computer Age," *Drew Gateway* 39, no. 3 (1969); G. Hinson, "Christianity in the Computer Age," *Review & Expositor* 62, no. 4 (1965); W. Gordon Ross, "Christianity in a Computer Age," *Religion in Life* 34, no. 4 (1965).

11. Siker, *Liquid Scripture*, 53.
12. Siker, *Liquid Scripture*, 60.
13. Andersen taught at Fuller Theological Seminary in the 1990s. "Biography of Francis Andersen," n.d., accessed September 28, 2016, http://www.aiarch.org.au/fellows; E. Ann Eyland, "Revelations from Word Counts," in *Perspectives on Language and Text: Essays and Poems in Honor of Francis I. Andersen's Sixtieth Birthday, July 28, 1985*, ed. Edward G. Newing and Edgar W. Conrad (Winona Lake, IN: Eisenbrauns, 1987).
14. "What Is The GRAMCORD Institute?," n.d., accessed October 3, 2016, http://www.gramcord.org/whatis.htm.
15. Though Kraft attended evangelical Wheaton College, he would later describe himself as a "progressive Christian" in a video. "Robert A. Kraft on Progressive Christianity," June 14, 2012, https://www.youtube.com/watch?v=F5UxVvUVRu0.
16. Hughes went on to edit and write for *Bits and Bytes Review*, a newsletter that advertised on the cover of his book and ran from 1986 to 1991.
17. Hughes, *Bits, Bytes and Biblical Studies* 344.
18. Darren Dochuk, *From Bible Belt to Sunbelt: Plain-Folk Religion, Grassroots Politics, and the Rise of Evangelical Conservatism* (New York: Norton, 2010).
19. Robert Wuthnow, *Rough Country: How Texas Became America's Most Powerful Bible-Belt State* (Princeton, NJ: Princeton University Press, 2014).
20. Additional details of Bible Research System's version history were provided to the author via email from Bert Brown in October 2016.
21. Gustin, "25th Anniversary for Bible Software Is Celebrated with New Release," October 11, 2005, https://www.businesswire.com/news/home/20051011005141/en/25th-Anniversary-Bible-Software-Celebrated-New-Release.
22. Andrew Pollack, "Putting Out the Word Electronically," *New York Times*, August 29 1982, http://www.nytimes.com/1982/08/29/weekinreview/putting-out-the-word-electronically.html.
23. Hunter, "How They Got the Bible on Disk."
24. Lubeck, "Scrolling through the Bible," 208.
25. John Edwards, "The Disk-Based Bible," 12, no. 2, *Popular Computing* (1983), 97, https://archive.org/stream/militarychaplain35unse/militarychaplain35unse_djvu.txt.
26. Personal correspondence with Bert Brown, September 28, 2016.
27. "Computer Software Can Aid Preachers," *Preaching*, 1987, http://beta.preaching.com/resources/articles/11566947/.
28. The italicized text was bolded in the original advertisement. "The SCRIPTURE SCANNER Advertisement," *PC Mag*, 1984.
29. Jeffrey Hsu, *A Comprehensive Guide to Computer Bible Study: Up-to-Date Information on the Best Software and Techniques* (Dallas: Word Publishing, 1993), 98.
30. Robert W. Klein, "Something New under the Sun: Computer Concordances and Biblical Study," *Christian Century* 114, no. 32 (1997).
31. The article included a header that said it was a "Special Advertising Section," meaning that at least some of the software applications paid to be included or written more about. Academic applications included BibleWorks, Accordance, Gramcord, and

Bible Windows (later renamed iBiblio). Professional applications (designed for pastors and other ministry leaders) included PC Study Bible, Logos, Epiphany, WORDsearch. Devotional applications included QuickVerse and a KJV application for Apple's portable Newton device.

32. Lori Hawkins, "Scriptures on Screen," *Austin American-Statesman*, December 10 1993, C1; "LifeWay's New Bible Software Debuts among SBC Seminarians," updated October 17, 2003, http://www.bpnews.net/16883/lifeways-new-bible-software-debuts-among-sbc-seminarians.
33. "WORDsearch and Epiphany Software Join Forces," WORDSearch, updated July 14, 2003, http://web.archive.org/web/20110718113235/http://www.wordsearchbible.com/about/pressreleases/pr-epiphany-wordsearch-merger.php; Jonathan Kever, "The Year's Best Software for Preachers: Version 2003," *Preaching*, 2003, accessed May 19, 2022, https://www.preaching.com/articles/the-years-best-software-for-preachers-version-2003/.
34. After the merger, the partnership with LifeWay continued, and WORDsearch would go on to produce Bible Navigator as LifeWay's main Bible application. "Ease of Use, Speed Linked in New LifeWay Bible Software," *Baptist Press*, updated October 17, 2003, http://www.bpnews.net/16884/ease-of-use-speed-linked-in-new-lifeway-bible-software.
35. "Wordsearch Bible Homepage," 2017, accessed January 23, 2017, https://www.wordsearchbible.com.
36. "Wordsearch Bible Is Transitioning to Logos," updated September 21, 2020, https://blog.faithlife.com/blog/2020/09/wordsearch-bible-is-transitioning-to-logos/.
37. Marilyn Deegan and Kathryn Sutherland, *Transferred Illusions: Digital Technology and the Forms of Print* (New York: Routledge, 2009).
38. Hsu listed CDWord under a special category of applications using CD-ROMs that held considerably more data than applications released through diskettes. Hsu, *Comprehensive Guide*, 206.
39. John D. Hannah, *An Uncommon Union: Dallas Theological Seminary and American Evangelicalism* (Grand Rapids, MI: Zondervan, 2009), 204.
40. Kenneth Alan Daughters, "Review of Computer Concordance Software," *Emmaus Journal* 2, no. 1 (1993): 88.
41. Michael Duduit, "Using Your Computer in Sermon Preparation," *Preaching*, September 1, 1990, 88, http://www.preaching.com/resources/articles/11567226/.
42. Harbin, "Fiat Lux," 28.
43. Keathley's father and Bob Pritchett's father were both DTS graduates. The detail about Keathley's role in the sale of CDWord to Logos was related to me by W. Hall Harris in a personal conversation February 22, 2017.
44. "About Libronix," 2006, accessed March 2, 2017, http://web.archive.org/web/20060101130140/http://www.libronix.com/page.aspx?id=about.
45. "Logos Bible Software Rebrands as Faithlife, Acquires Beacon Ads," Logos, updated October 7, 2014, https://web.archive.org/web/20160304091759/https://www.logos.com/press/releases/Logos-Bible-Software-rebrands-as-Faithlife-acquires-Beacon-Ads.

46. Daughters, "Review of Computer Concordance Software," 87.
47. Kenneth Alan Daughters, "Logos Bible Software 2.0 Level 4," *Emmaus Journal* 4, no. 1 (1995).
48. David Parker, "Logos Bible Software Series X Scholar's Library Biblical Languages Supplement," *Evangelical Review of Theology* 27, no. 3 (2003); Tim Challies, "Logos Bible Software 3," *Journal of Modern Ministry* 3, no. 3 (Fall 2006), https://www.logos.com/press/reviews/challies; "Logos Bible Software 3—Scholar's Library: Gold," Bible Software Review, updated June 2, 2007, http://web.archive.org/web/20111011084122/http://www.bsreview.org/index.php?modulo=Reviews&id=10.
49. A reflexive note: In 2013, while working at Dallas Theological Seminary, I was part of a team that evaluated Bible software that DTS would implement for all its students. While DTS's New Testament faculty tended to prefer BibleWorks and the Old Testament faculty used Accordance, the faculty agreed that Logos would be a better overall package because it was available on more platforms (PC, Mac, iPhone, Android) and had a greater variety of resources available for students to purchase.
50. "Faithlife CEO Bob Pritchett on Balancing the Principles of Faith, Science and Entrepreneurship," Beliefnet, 2015, accessed February 10, 2017; "40 Under 40 Alumni," *Puget Sound Business Journal*, 2015, accessed June 10, 2016, http://www.bizjournals.com/seattle/special/2015/40under40seattle/Alumni; "Faithlife's Bob Pritchett Ranked among Top 25 CEOs in 2015," Logos, updated June 29, 2015, https://blog.faithlife.com/blog/2015/06/faithlifes-bob-pritchett-among-top-25-ceos-of-2015/.
51. United States Census Bureau, *Home Computers and Internet Use in the United States: August 2000*, United States Census Bureau, September 2001, https://www.census.gov/prod/2001pubs/p23-207.pdf.
52. Nick Hengeveld, "[comp.infosystems.www] Anyone want to test a gateway?," updated December 28, 1993, https://groups.google.com/forum/#!msg/comp.archives/qDBF4nb7jSk/AxgqJK6BsusJ.
53. Rachel Barach, "Celebrating an Online Bible Legacy: Bible Gateway," accessed June 12, 2018, https://www.biblegateway.com/article.
54. "About Bible Gateway," accessed February 22, 2017, https://www.biblegateway.com/about/
55. Interviews conducted December 3, 2014.
56. Nick Hengeveld, LinkedIn, Retrieved February 22, 2017, from https://www.linkedin.com/in/nickhengeveld.
57. "Futures for the Bible Gateway," 1997, accessed December 10, 2017, http://web.archive.org/web/19971210071327/http://bible.gospelcom.net/bg/futures.html.
58. "Bible Gateway homepage," 1998, accessed January 12, 2018, http://web.archive.org/web/19980110131629/http://bible.gospelcom.net/.
59. Clayton Hardiman, "Gospel Communications Online Sold," *Muskegon Chronicle*, October 29 2008, http://blog.mlive.com/chronicle/2008/10/gospel_communications_online_s.html. Julia Bauer, "Zondervan Acquires Religious Site BibleGateway.com," *Grand Rapids Press*, October 28, 2008, http://www.mlive.com/grpress/business/index.ssf/2008/10/zondervan_acquires_religious_s.html.

60. Weber, "Why Zondervan Bought BibleGateway.com."
61. "On 25th Anniversary, Bible Gateway Has Been Viewed More Than 14 Billion Times," Bible Gateway, updated August 1, 2018, https://www.biblegateway.com/blog/2018/08/on-25th-anniversary-bible-gateway-has-been-viewed-more-than-14-billion-times/.
62. Phone interview with J. Hampton Keathley IV on February 22, 2017.
63. "Bibles," January 10, 1998, http://web.archive.org/web/19980110133428/http://bible.com/bibles.html.
64. "SCRIPTURES.com Homepage," October 13, 1999, http://web.archive.org/web/19991013082020/http://scriptures.com/.
65. "Read the Bible Online!," February 29, 2000, http://web.archive.org/web/20000229202117/http://www.bible.com/bible_read.html.
66. "JAMES R. SOLAKIAN vs. BIBLE.COM, INC.," updated October 19, 2010, accessed December 12, 2017, https://picker.typepad.com/files/bible.com.pdf.
67. Tom Hals, "Bible.com Investor Sues Company for Lack of Profit," *Reuters*, October 21, 2010, http://www.reuters.com/article/us-biblecom-lawsuit-idUSTRE69K42D20101021.
68. "Bible.com joins the YouVersion family!," *YouVersion* blog, updated August 30, 2012, http://blog.youversion.com/2012/08/bible-com/.
69. The dates and details of the formation of the Biblical Studies Foundation and Biblical Studies Press were provided by a phone interview with J. Hampton Keathley on February 22, 2017.
70. "Biblical Studies Foundation Homepage," October 30, 1996, http://web.archive.org/web/19961030041731/http://www.bible.org/.
71. "J. Hampton Keathley, III, Author Page," accessed February 20, 2017, https://bible.org/users/j-hampton-keathley-iii.
72. "About Us," n.d., accessed February 20, 2017, https://bible.org/book/about-bibleorg.
73. Daniel B. Wallace, "Innovations in Text and Translation of the NET Bible, New Testament," *Bible Translator* 52, no. 3 (2001). Again, for full disclosure, I mention that as of this writing, I am a graduate and current employee of Dallas Theological Seminary, although I did not have any involvement with the formation of the NET Bible. I was studying for my driver's license examination at the time.
74. Daniel B. Wallace, "An Open Letter Regarding the NET Bible, New Testament," *Notes on Translation* 14, no. 3 (2000).
75. Fackler, "Second Coming of Holy Writ."
76. "The New English Translation (NET) Bible Available in Print with 60,000 Translators' Notes," Bible Gateway, updated August 14, 2019, https://www.biblegateway.com/blog/2019/08/the-new-english-translation-net-bible-available-in-print-with-60000-translators-notes/.
77. "Conspicuous Consumption and Your iPhone," 2008, accessed March 17, 2017, http://www.everydaysociologyblog.com/2008/09/conspicuous-con.html; Katja Rakow, "The Bible in the Digital Age: Negotiating the Limits of 'Bibleness' of Different Bible Media," in *Christianity and the Limits of Materiality*, ed. Minna Opas and Anna Haapalainen (London: Bloomsbury, 2016).

78. Alexander Chow, "What Has Jerusalem to Do with the Internet? World Christianity and Digital Culture," *International Bulletin of Mission Research* (2021), https://www.research.ed.ac.uk/en/publications/what-has-jerusalem-to-do-with-the-internet-world-christianity-and.
79. Mentioned in Steve Deyo, "Cyber-Boost Your Faith: The Latest Bible Study Software Will Empower Your Study at Any Level," *Christianity Today*, 42, no. 5 (1998), 16, EBSCOhost.
80. "About Olive Tree Bible Software," Olive Tree, https://www.olivetree.com/press/, accessed March 5, 2017; "Brief History of Olive Tree," Olive Tree, https://www.olivetree.com/press/history.php, accessed March 5, 2017.
81. Lynn Garrett, "Harper Acquires Bible Software Company," *Publishers Weekly*, May 6, 2014, http://www.publishersweekly.com/pw/by-topic/industry-news/religion/article/62160-harper-acquires-bible-software-company.html.
82. "Mobilize Your Bible Study with Olive Tree," http://animatedfaith.com/olive_tree_bible_software.php, accessed March 5, 2017.
83. Kevin Purcell, "Interview with Stephen Johnson, CEO of OliveTree Bible Software," *Church Tech Today*, September 26, 2012, http://churchtechtoday.com/2012/09/26/interview-stephen-johnson-ceo-olivetree-bible-software/.
84. "About Laridian," https://www.laridian.com/content/about.asp, accessed March 17, 2017.
85. M. Dubis, "The Bible in Your Palm: Biblical Studies Software for Palm OS Handheld Devices," *Journal of Religious & Theological Information* 4, no. 4 (2001); Tim Gnatek, "A New Bible, Palmtop Version, Can Keep Track of Studies," *New York Times*, December 2 2004, http://www.nytimes.com/2004/12/02/technology/circuits/a-new-bible-palmtop-version-can-keep-track-of-studies.html.
86. "Mobile Fact Sheet," Pew Reseach Center, April 7, 2017, http://www.pewinternet.org/fact-sheet/mobile/.
87. "YouVersion Homepage," 2007, accessed June 15, 2017, http://web.archive.org/web/20070705040140/http://www.youversion.com/
88. Tim Hutchings, "Creating Church Online an Ethnographic Study of Five Internet-Based Christian Communities" (PhD diss., Durham University, 2010).
89. "My Take: How Technology Could Bring Down the Church," CNN, updated May 15, 2011, http://religion.blogs.cnn.com/2011/05/15/my-take-how-technology-could-bring-down-the-church/comment-page-8/.
90. Hutchings, "Bible Is an App."
91. Terry Storch, "YouVersion Now Available on iPhone," July 11, 2008, http://web.archive.org/web/20080718040045/http://blog.youversion.com/?p=97.
92. Hutchings, "Bible Is an App."
93. "With 100 Million Downloads, YouVersion Bible Is a Massive App That No VC Can Touch," *Business Insider*, updated July 29, 2013, https://www.businessinsider.com/youversion-bible-app-has-100-million-downloads-2013-7.
94. Brian Solomon, "Meet David Green: Hobby Lobby's Biblical Billionaire," *Forbes*, September 18, 2012, https://www.forbes.com/sites/briansolomon/2012/09/18/david-green-the-biblical-billionaire-backing-the-evangelical-movement/.

Chapter 5

1. Campbell, *Digital Creatives*.
2. My personal relationships figured heavily in the access I was given at each company. This included Rachel Barach, general manager of Bible Gateway, and Stephen Smith, senior director of digital products for Bible Gateway, as well as Bob and Dan Pritchett of Faithlife, whom I met through my job at Dallas Theological Seminary. I had more trouble connecting with the YouVersion team, but was eventually able to reach them through a friend.
3. Campbell, *When Religion Meets New Media*, 60–61.
4. Morgan and Liker outline the sociotechnical system at Toyota using this outline. This approach is also followed by the American television program *The Profit*, in which entrepreneur Marcus Lemonis attempts to improve companies by focusing on people, product, and process. James M. Morgan and Jeffrey K. Liker, *The Toyota Product Development System: Integrating People, Process, and Technology* (London: Taylor & Francis, 2006).
5. The names of all the individuals I interviewed have been removed. I do refer to some individuals as "developers" or "leaders" to distinguish their roles when it contributes to understanding their statements, but I have attempted to avoid including any personally identifying information.
6. I wrote about some of my early experiences with technology in faith in John Dyer, *From the Garden to the City: The Redeeming and Corrupting Power of Technology* (Grand Rapids, MI: Kregel, 2011).
7. James Matthew Price, "Undergraduate Perceptions of Vocational Calling into Missions and Ministry," *Missiology* 41, no. 1 (2013), https://doi.org/10.1177/0091829612466997.
8. Abby Day, *Believing in Belonging: Belief and Social Identity in the Modern World* (Oxford: Oxford University Press, 2011).
9. Robert Wuthnow, "Taking Talk Seriously: Religious Discourse as Social Practice," *Journal for the Scientific Study of Religion* 50, no. 1 (2011): 9.
10. Malan Nel and Eric Scholtz, "Calling, Is There Anything Special about It?," *Hervormde Teologiese Studies* 72, no. 4 (2016), https://doi.org/10.4102/hts.v72i4.3183. They outline the historical use of "calling," particularly within the Presbyterian tradition, while Richard N. Pitt, *Divine Callings: Understanding the Call to Ministry in Black Pentecostalism* (New York: New York University Press, 2012), explores it in black Pentecostalism.
11. In an example of popular-level writing on the subject, Smith speaks of three callings: (1) as a Christian, (2) as "a defining purpose or mission, a reason for being" and (3) as our immediate, daily duties. Gordon T. Smith, *Courage and Calling: Embracing Your God-Given Potential*, rev. ed. (Downers Grove, Ill: IVP Books, 2011), 9–10.
12. Selected examples include Timothy Keller and Katherine Leary Alsdorf, *Every Good Endeavour: Connecting Your Work to God's Plan for the World* (London: Hodder & Stoughton, 2014); Amy L. Sherman, *Kingdom Calling: Vocational Stewardship for the Common Good* (Downers Grove, IL: IVP Books, 2011); and Ben Witherington III, *Work: A Kingdom Perspective on Labor* (Grand Rapids, MI: Eerdmans, 2011).

13. Nick Bilton, "Steve Jobs Was a Low-Tech Parent," *New York Times*, September 10, 2014, https://www.nytimes.com/2014/09/11/fashion/steve-jobs-apple-was-a-low-tech-parent.html; "Billionaire Tech Mogul Bill Gates Reveals He Banned His Children from Mobile Phones until They Turned 14," *The Mirror*, updated April 20, 2017, http://www.mirror.co.uk/tech/billionaire-tech-mogul-bill-gates-10265298. Bowles Nellie Bowles, "The Digital Gap between Rich and Poor Kids Is Not What We Expected," *New York Times*, October 26 2018, https://www.nytimes.com/2018/10/26/style/digital-divide-screens-schools.html, argues that the new digital divide is between students forced to use tablets and those who can afford to attend schools that still have physical books, blocks, and other learning opportunities.
14. David Sax, *The Revenge of Analog: Real Things and Why They Matter* (New York City: PublicAffairs, 2016).
15. Taylor University, 2011 #657.
16. Pamela Caudill Ovwigho and Arnold Cole, "Scriptural Engagement, Communication with God, and Moral Behavior among Children," *International Journal of Children's Spirituality* 15, no. 2 (2010), https://doi.org/10.1080/1364436X.2010.497642; Pamela Caudill Ovwigho and Arnold Cole, "Understanding the Bible Engagement Challenge: Scientific Evidence for the Power of 4," 2009, Center of Bible Engagement, https://web.archive.org/web/20161215103917/http://www.backtothebible.org/files/web/docs/cbe/Scientific_Evidence_for_the_Power_of_4.pdf.
17. Hiemstra, *Confidence Conversation and Community*, 8.
18. T. Wayne Dye, "Scripture in an Accessible Form: The Most Common Avenue to Increased Scripture Engagement," *International Journal of Frontier Missiology* 26, no. 3 (2009); T. Wayne Dye, "The Eight Conditions of Scripture Engagement: Social and Cultural Factors Necessary for Vernacular Bible Translation to Achieve Maximum Effect," *International Journal of Frontier Missiology* 26, no. 2 (2009).
19. Malley, *How the Bible Works*, 105.
20. Bielo, "Failure of Meaning."
21. Hutchings, "Studying Apps," 62.
22. "HarperCollins Christian Publishing Acquires Olive Tree," HarperCollins, updated May 5 2014, https://www.harpercollinschristian.com/harpercollins-christian-publishing-acquires-olive-tree/.
23. "Introducing the Every Day Bible App," updated December 3, 2014, https://blog.faithlife.com/blog/2014/12/introducing-the-every-day-bible-app/.
24. Fulk, "Social Construction of Communication Technology," 922. Malcolm Torry, *Managing God's Business: Religious and Faith-Based Organizations and Their Management* (New York: Routledge, 2016), 8.
25. Eric von Hippel, *The Sources of Innovation* (Oxford: Oxford University Press, 1998).
26. For example, the progressive Our Bible app mentioned in chapters 1 and 4 has devotionals entitled "Asexuali-TEA" and "(S)Exodus: Affirming Your Identity, Body, and Fight for Justice Through Exodus" that do not appear in the YouVersion devotional list. Retrieved March 7, 2019.
27. See Crouch's five orientations toward culture, mentioned in chapter 3, one of which is "copying." Crouch, *Culture Making*, 78–100.

28. "Why Friendships in Bible App 5 Are Different," YouVersion, updated April 8, 2014, http://blog.youversion.com/2014/04/why-friendships-in-bible-app-5-are-different/.
29. Elizabeth Day, "How Selfies Became a Global Phenomenon," *The Guardian*, July 13, 2013, http://www.theguardian.com/technology/2013/jul/14/how-selfies-became-a-global-phenomenon; Molly Wood, "Narcissist's Dream: Selfie-Friendly Phone," *New York Times*, February 6, 2014, https://www.nytimes.com/2014/02/06/technology/personaltech/making-the-case-for-a-more-selfie-friendly-smartphone.html.
30. "The Origin of the Retweet and Other Twitter Arcana," Medium, updated March 6, 2019, https://medium.com/@narendra/the-origin-of-the-retweet-and-other-twitter-arcana-5c53289d9a47; "The First-Ever Hashtag, @-reply and Retweet, as Twitter Users Invented Them," Quartz, updated October 15, 2013, https://qz.com/135149/the-first-ever-hashtag-reply-and-retweet-as-twitter-users-invented-them/.
31. Companies like Facebook and Google are regularly in the news as users and advocacy groups express privacy concerns about the way they use data. Carole Cadwalladr and Emma Graham-Harrison, "Revealed: 50 Million Facebook Profiles Harvested for Cambridge Analytica in Major Data Breach," *The Guardian*, March 17, 2018, https://www.theguardian.com/news/2018/mar/17/cambridge-analytica-facebook-influence-us-election; "Google's Gmail Controversy Is Everything People Hate about Silicon Valley," updated July 3, 2018, https://www.cnet.com/news/privacy/googles-gmail-controversy-is-everything-wrong-with-silicon-valley/.
32. Douglas, "Smooth Scrolling," Faithlife Forums, October 1, 2011, https://community.logos.com/forums/p/38920/338605.aspx.
33. Eric Vaniman, "Select Scrolling Options in Your Logos Bible App," Logos Talk Blog, March 19, 2015. https://web.archive.org/web/20150919235942/https://blog.logos.com/2015/03/select-scrolling-options-in-your-logos-bible-app/.
34. Peter M. Phillips, *The Bible, Social Media and Digital Culture* (London: Routledge, 2019).
35. "The 100 Most Read Bible Verses at Bible Gateway," Bible Gateway, updated May 15, 2009, https://www.biblegateway.com/blog/2009/05/the-100-most-read-bible-verses-at-biblegatewaycom/; "This Is the Most Popular Verse in 2 Billion Pageviews during 2018 on Bible Gateway," Bible Gateway, updated December 10, 2018, https://www.biblegateway.com/blog/2018/12/this-is-the-most-popular-verse-in-2-billion-pageviews-during-2018-on-bible-gateway/.
36. Designer Sebastian Lindemann's work serves as an example of this design discussion. "Bye, Bye Burger!," updated October 11, 2014, https://medium.com/startup-grind/bye-bye-burger-5bd963806015.
37. John H. Hayes, "The Evangelical Ethos and the Spirit of Capitalism," *Perspectives in Religious Studies* 39, no. 3 (2012).
38. Connolly also suggests that evangelicalism has a sense of resentment and desire for revenge that maps to capitalism's sense of entitlement and disregard for others. However, this sentiment was not detected in the interviews, and some of the interviewees seemed more likely to fit into Connolly's "dissidents on the edge of the machine." Connolly, "Evangelical-Capitalist Resonance Machine," 883. In particular, comments from the female executive in this study seemed to follow Ferguson's argument that evangelical women do not conform to the same patterns. Kathy E.

Ferguson, "Bringing Gender into the Evangelical-Capitalist Resonance Machine," *Political Theology* 12, no. 2 (2011), https://doi.org/10.1558/poth.v12i2.184.

39. Bible Gateway also offers users the chance to pay for access to additional books and a monthly payment plan to access resources on demand.
40. Candida R. Moss and Joel S. Baden, *Bible Nation: The United States of Hobby Lobby* (Princeton, NJ: Princeton University Press, 2017).
41. YouVersion, "Easter Bible Engagement"; YouVersion, "Infographics"; "YouVersion Expects Record-Breaking Bible Plan Completions This Easter," updated April 8, 2019, https://www.youversion.com/press/youversion-expects-record-breaking-bible-plan-completions-this-easter/.
42. Bill Leonard, "Getting Saved in America: Conversion Event in a Pluralistic Culture," *Review & Expositor* 82, no. 1 (Winter 1985): 113.
43. In his presentation, Donaldson also invoked the parable of the talents. "Faith Leads Tech," Nashville, Tennessee, November 9, 2018, http://faithleads.tech/. I also presented at this conference.

Chapter 6

1. Some of the data in this and the following chapter were incorporated in a journal article. John Dyer, "The Habits and Hermeneutics of Digital Bible Readers: Comparing Print and Screen Engagement, Comprehension, and Behavior," *Journal of Religion, Media and Digital Culture* 8, no. 2 (2019), https://doi.org/https://doi.org/10.1163/21659214-00802001, https://brill.com/view/journals/rmdc/8/2/article-p181_181.xml.
2. Though Hutchby posits his work in contrast to social shaping of technology, which he sees as too focused on the social, he, too, sought to chart a "third way" between determinism and instrumentalism. Ian Hutchby, "Technologies, Texts and Affordances," *Sociology* 35, no. 2 (2001), https://doi.org/10.1017/S0038038501000219.
3. Bob Smietana, "Bible App Offers Portability for the Faithful," *USA Today*, August 13 2013, https://www.usatoday.com/story/tech/2013/08/11/youversion-bible-app/2640719/.
4. As discussed in chapter 2, I chose Jude for the comprehension assessment because it takes the familiar form of an epistle but is not often studied or preached on. I chose the Gospel of John for the reading plan because it is a familiar text and because YouVersion had a ten-day reading plan of the book.
5. The names of the churches are pseudonyms.
6. Some did not sign the consent waiver, while others signed it but did not fill out the survey questions or complete the comprehension assessment.
7. Barna Group, *The Bible in America: The Changing Landscape of Perceptions and Engagement* (Ventura, CA: Barna Group, 2016), 71.
8. American Bible Society, *The State of the Bible 2016*, 2016, accessed May 20, 2022, http://www.americanbible.org/uploads/content/State_of_the_Bible_2016_report_Politics.pdf.

9. As mentioned previously, Malley found that many participants reported that their reading had only recently dropped off. Malley, *How the Bible Works*, 35.
10. David G. Ford, Joshua L. Mann, and Peter M. Phillips, *The Bible and Digital Millennials* (New York: Routledge, 2019), 68, 73.
11. Rakow, "Bible in the Digital Age."
12. Malley documented the importance of a "first Bible" in faith formation. Malley, *How the Bible Works*, 67–68.
13. Pew Research Center, *Technology Device Ownership: 2015*, October 2015, http://www.pewinternet.org/2015/10/29/technology-device-ownership-2015.
14. Tim Hutchings, "E-reading and the Christian Bible," *Studies in Religion/Studies Religieuses* 44, no. 4 (October 16, 2015): 437.
15. Evan Howard, "*Lectio Divina* in the Evangelical Tradition," *Journal of Spiritual Formation & Soul Care* 5, no. 1 (2012).
16. A few participants wrote that they used CDs or other media not included in the survey to listen to audio Bibles.
17. Malley works through the various social constructions and meanings of the phrase "what the Bible says about" and its relationship to evangelical biblicism.Brian Malley, "Understanding the Bible's Influence," in *The Social Life of Scriptures; Cross-Cultural Perspectives on Biblicism*, ed. James Bielo (New Brunswick, NJ: Rutgers, 2009).
18. James S. Bielo, "Textual Ideology, Textual Practice: Evangelical Bible Reading in Group Study," in *The Social Life of Scriptures; Cross-Cultural Perspectives on Biblicism*, ed. James S. Bielo (New Brunswick, NJ: Rutgers, 2009), 173.
19. Bielo also argues that while accurately interpreting scripture is important to evangelical Bible studies, the process of establishing the Bible's "authority, relevance, and textuality" is perhaps more critical for small-group interaction. Bielo, *Words upon the Word*, 70.
20. "What's Good about Having the Bible on Mobile Devices? What's Dangerous?," Brookside Institute, updated September 3, 2015, https://www.thebrooksideinstitute.net/blog/whats-good-about-having-the-bible-on-mobile-devices-whats-dangerous.
21. James S. Bielo, "Cultivating Intimacy: Interactive Frames for Evangelical Bible Study," *Fieldwork in Religion* 3, no. 1 (2008): 62.
22. Kathleen C. Boone, *The Bible Tells Them So: The Discourse of Protestant Fundamentalism* (Albany: State University of New York Press, 1989), 40.
23. Campbell, "Understanding the Relationship," 74–76.
24. The concept of the perspicuity of scripture dates back to the Reformers who contrasted it with the need for official church interpretation. Mickey L. Mattox, "Martin Luther," in *Christian Theologies of Scripture: A Comparative Introduction*, ed. Justin Holcomb (New York: New York University Press, 2006), 104. Evangelicals like Vanhoozer continue to argue for a form of perspicuity, but scholars like Stackhouse argue that it is often practiced in "in ways that are sometimes perplexing and even self-defeating." Kevin J. Vanhoozer, *The Drama of Doctrine: A Canonical Linguistic Approach to Christian Doctrine* (Louisville: Westminster John Knox Press, 2005), 206[John G. Stackhouse Jr., "Evangelicals and the Bible Yesterday, Today, and Tomorrow," in *New Paradigms for Bible Study: The Bible in the Third Millennium*, ed. Robert M. Fowler, Edith Blumhofer, and Fernando F. Segovia (New York: Continuum, 2004), 188.

25. "Westminster Confession of Faith," 1646, accessed May 20, 2022, http://www.ccel.org/ccel/anonymous/westminster3.i.i.html.
26. Ryan Torma and Paul Emerson Teusner, "iReligion," *Studies in World Christianity* 17, no. 2 (2011), https://doi.org/10.3366/swc.2011.0017; J. O'Neill, "Digital Annotation: Not There Yet," *Information Today* 27, no. 7 (July 2010).
27. Noll, *American Evangelical Christianity*, 2.
28. Phillips, *Bible, Social Media*, introduction.
29. Phillips, *Bible, Social Media*, chapter 2.
30. Eyal Ophira, Clifford Nass, and Anthony D. Wagner, "Cognitive Control in Media Multitaskers," *Proceedings of the National Academy of Sciences* 106, no. 33 (2009); Melina R. Uncapher and Anthony D. Wagner, "Minds and Brains of Media Multitaskers: Current Findings and Future Directions," *Proceedings of the National Academy of Sciences* 115, no. 40 (2018), https://doi.org/10.1073/pnas.1611612115, http://www.pnas.org/content/115/40/9889.abstract; Yvonne Ellis, Bobbie Daniels, and Andres Jauregui, "The Effect of Multitasking on the Grade Performance of Business Students," *Research in Higher Education Journal* 8, no. 1 (2010); Daniel J. Levitin, *The Organized Mind: Thinking Straight in the Age of Information Overload* (Boston: Dutton, 2015), 97; Abraham Flanigan and Markeya Peteranetz, "Digital Distraction across Courses: Self-Regulation of Digital Device Use in Favorite versus Least Favorite Courses," paper presented at the American Educational Research Association, Annual Meeting, Toronto, April 5–9, 2019.
31. Eric H. Schumacher et al., "Virtually Perfect Time Sharing in Dual-Task Performance: Uncorking the Central Cognitive Bottleneck," *Psychological Science* 12, no. 2 (March 1, 2001), https://doi.org/10.1111/1467-9280.00318, https://doi.org/10.1111/1467-9280.00318.
32. YouVersion, "The Bible App for Kids Is Here!," April 12, 2013, https://blog.youversion.com/2013/11/the-bible-app-for-kids-is-here/.
33. Matt Richtel, "Digital Devices Deprive Brain of Needed Downtime," *New York Times*, August 24 2010, https://www.nytimes.com/2010/08/25/technology/25brain.html.
34. Siker, *Liquid Scripture*, 185.
35. Ericsson, *Mobility Report*, June 3, 2018, https://www.ericsson.com/en/mobility-report.
36. Campbell, "Understanding the Relationship."
37. Naomi S. Baron, *Words Onscreen: The Fate of Reading in a Digital World* (New York: Oxford University Press, 2015).
38. Howard Gardner and Katie Davis, *The App Generation: How Today's Youth Navigate Identity, Intimacy, and Imagination in a Digital World* (New Haven: Yale University Press, 2013); Lee Rainie and Barry Wellman, *Networked: The New Social Operating System* (Cambridge, MA: MIT Press, 2012).

Chapter 7

1. Maryanne Wolf and Catherine J. Stoodley, *Proust and the Squid: The Story and Science of the Reading Brain* (New York: HarperCollins, 2007). Dehaene makes a similar

argument, drawing on neuroscience suggesting that the human brain did not evolve for the task of reading, though he does not offer deep consideration of the influence of screens. Stanislas Dehaene, *Reading in the Brain: The Science and Evolution of a Human Invention* (New York: Viking, 2009).

2. Maryanne Wolf and C. J. Stoodley, *Reader, Come Home: The Reading Brain in a Digital World* (New York: Harper, 2018), 35ff.
3. Andrew Dillon, "Reading from Paper versus Screens: A Critical Review of the Empirical Literature," *Ergonomics* 35, no. 10 (1992).
4. W. H. Cushman, "Reading from Microfiche, VDT and the Printed Page: Subjective Fatigue and Performance," *Human Factors* 28, no. 1 (1986).
5. Erik Wästlund et al., "Effects of VDT and Paper Presentation on Consumption and Production of Information: Psychological and Physiological Factors," *Computers in Human Behavior* 21, no. 2 (2005), https://doi.org/10.1016/j.chb.2004.02.007.
6. J. M. Noyes and K. J. Garland, "Computer- vs. Paper-Based Tasks: Are They Equivalent?," *Ergonomics* 51, no. 9 (September 2008), https://doi.org/10.1080/00140130802170387, http://www.ncbi.nlm.nih.gov/pubmed/18802819.
7. Kerstin Severinson Eklundh, "Problems in Achieving a Global Perspective of the Text in Computer-Based Writing," *Instructional Science* 21, no. 1 (1992); Annie Piolat, Jean-Yves Roussey, and Olivier Thunin, "Effects of Screen Presentation on Text Reading and Revising," *International Journal of Human-Computer Studies* 47 (1996); T. Baccino, *Lecture électronique: De la vision à la compréhension* (Grenoble: Presses Universitaires de Grenoble, 2004).
8. Diana DeStefano and Jo-Anne LeFevre, "Cognitive Load in Hypertext Reading: A Review," *Computers in Human Behavior* 23, no. 3 (2007).
9. Anne Mangen, "Hypertext Fiction Reading: Haptics and Immersion," *Journal of Research in Reading* 31, no. 4 (2008), http://dx.doi.org/10.1111/j.1467-9817.2008.00380.x; Ziming Liu, "Reading Behaviour in the Digital Environment: Changes in Reading Behaviour over the Past Ten Years," *Journal of Documentation* 61, no. 6 (2005), http://dx.doi.org/10.1108/00220410510632040.
10. Terje Hillesund, "Digital Reading Spaces: How Expert Readers Handle Books, the Web and Electronic Paper," *First Monday* 15, nos. 4–5 (2010), https://doi.org/10.5210/fm.v15i4.2762 .
11. Anne Mangen, Bente R. Walgermo, and Kolbjørn Brønnick, "Reading Linear Texts on Paper versus Computer Screen: Effects on Reading Comprehension," *International Journal of Educational Research* 58 (2013).
12. Guang Chen et al., "A Comparison of Reading Comprehension across Paper, Computer Screens, and Tablets: Does Tablet Familiarity Matter?," *Journal of Computers in Education* 1, nos. 2–3 (2014).
13. Rakefet Ackerman and Tirza Lauterman, "Taking Reading Comprehension Exams on Screen or on Paper? A Metacognitive Analysis of Learning Texts under Time Pressure," *Computers in Human Behavior* 28, no. 5 (2012), https://doi.org/10.1016/j.chb.2012.04.023; Rakefet Ackerman and Morris Goldsmith, "Metacognitive Regulation of Text Learning: On Screen versus on Paper," *Journal of Experimental Psychology: Applied* 17, no. 1 (2011), http://psycnet.apa.org/journals/xap/17/1/18/.

14. Anne Mangen et al., "Mystery Story Reading in Pocket Print Book and on Kindle: Possible Impact on Chronological Events Memory," paper presented at a conference of the International Society for the Empirical Study of Literature and Media, Turin, Italy, July 21–25, 2014.
15. Szu-Yuan Sun, Chich-Jen Shieh, and Kai-Ping Huang, "A Research on Comprehension Differences between Print and Screen Reading," *South African Journal of Economic and Management Sciences* 16 (2013), http://www.scielo.org.za/scielo.php?script=sci_arttext&pid=S2222-34362013000500010&nrm=iso.
16. Ziming Liu, *Paper to Digital: Documents in the Information Age* (Westport, CT: Libraries Unlimited, 2008); Åse Tveit and Anne Mangen, "A Joker in the Class: Teenage Readers' Attitudes and Preferences to Reading on Different Devices," *Library & Information Science Research* 35, nos. 3–4 (2014).
17. LCD screens have a backlight that can strain the eyes of the reader, where e-ink requires an external lighting source like a printed page. S. Benedetto et al., "E-readers and Visual Fatigue," *PLoS One* 8, no. 12 (2013), https://doi.org/doi:10.1371/journal.pone.0083676; Kaveri Subrahmanyam et al., "Learning from Paper, Learning from Screens," *International Journal of Cyber Behavior, Psychology and Learning* 3, no. 4 (2013), https://doi.org/10.4018/ijcbpl.2013100101; Sara J. Margolin et al., "E-readers, Computer Screens, or Paper: Does Reading Comprehension Change across Media Platforms?," *Applied Cognitive Psychology* 27, no. 4 (2013), https://doi.org/10.1002/acp.2930.
18. Assuming a benchmark of $p = .05$ as a measure of significance.
19. American Bible Society, *The State of the Bible 2019*; Sarah Eekhoff Zylstra, "What the Latest Bible Research Reveals about Millennials," *Christianity Today*, May 16, 2016, https://www.christianitytoday.com/news/2016/may/what-latest-bible-research-reveals-about-millennials.html.
20. Phillips, *Bible, Social Media.*
21. Philip Goff, Arthur E. Farnsley II, and Peter J. Thuesen, "The Bible in American Life Today," in *The Bible in American Life*, ed. Philip Goff, Arthur E. Farnsley II, and Peter J. Thuesen (Oxford: Oxford University Press, 2017), 17; Philip S. Brenner, "Cross-National Trends in Religious Service Attendance," *Public Opinion Quarterly* 80, no. 2 (Summer 2016), https://doi.org/10.1093/poq/nfw016, https://www.ncbi.nlm.nih.gov/pubmed/27274579 https://www.ncbi.nlm.nih.gov/pmc/PMC4888582/.
22. Amanda Friesen, "How American Men and Women Read the Bible," in *The Bible in American Life*, ed. Philip Goff, Arthur E. Farnsley II, and Peter J. Thuesen (Oxford: Oxford University Press, 2017).
23. Tim Hutchings, "'The Smartest Way to Study the Word': Protestant and Catholic Approaches to the Digital Bible," in *Negotiating Religious Visibility in Digital Media*, ed. Miriam Diez Bosch, Josep Lluís Micó, and Josep Maria Carbonell (Barcelona: Facultat de Comunicació i Relacions Internacionals Blanquerna, 2015).
24. Sarah Logan and Rhona Johnston, "Investigating Gender Differences in Reading," *Educational Review* 62, no. 2 (June 16, 2010): 175.
25. David Reilly, David Neumann, and Glenda Andrews, "Gender Differences in Reading and Writing Achievement: Evidence from the National Assessment of Educational Progress (NAEP)," *American Psychologist* 74, no. 4 (2019): 446–47.

26. Sun, Shieh, and Huang, "Research on Comprehension Differences."
27. Marta Trzebiatowska and Steve Bruce, *Why Are Women More Religious Than Men?* (Oxford: Oxford University Press, 2012).
28. Pew Research Center, *Religious Landscape Study* (2018), http://www.pewforum.org/religious-landscape-study/gender-composition/.
29. Women were also more likely to say that the most important biblical command is "Do justice, love mercy and walk humbly with God" (51% female, 49% male) while men were more likely to choose "Go and make disciples" (11% male, 5% female). American Bible Society, *The State of the Bible 2019*, 59, 64, 79, 84–47.
30. Hutchings, "E-reading," 432.
31. Lily Shashaani and Ashmad Khalili, "Gender and Computers: Similarities and Differences in Iranian College Students' Attitudes toward Computers," *Computers & Education* 37, no. 3 (November 1, 2001), https://doi.org/https://doi.org/10.1016/S0360-1315(01)00059-8.
32. Muriel Niederle and Lise Vesterlund, "Do Women Shy Away from Competition? Do Men Compete Too Much?," *Quarterly Journal of Economics* 122, no. 3 (2007), https://doi.org/10.1162/qjec.122.3.1067, https://doi.org/10.1162/qjec.122.3.1067.
33. C. Clark, "Redefining or Undermining? The Role of Technology in the Reading Lives of Children and Young People: Findings from the National Literacy Trust's Annual Survey 2012," National Literacy Trust (London, 2012).
34. Kate Summers, "Adult Reading Habits and Preferences in Relation to Gender Differences," *Reference and User Services Quarterly* 52, no. 3 (2013): 247–48.
35. Bureau of Labor Statistics, *American Time Use Survey—2014 Results*, US Department of Labor, June 24, 2015, https://www.bls.gov/news.release/archives/atus_06242015.pdf.
36. Thomas H. Davenport and John C. Beck, *The Attention Economy: Understanding the New Currency of Business* (Boston: Harvard Business School Press, 2001), 20.
37. Bianca Bosker, "The Binge Breaker," *The Atlantic*, November 2016, https://www.theatlantic.com/magazine/archive/2016/11/the-binge-breaker/501122/.
38. Justin Whitmel Earley, *The Common Rule: Habits of Purpose for an Age of Distraction* (Downers Grove, IL: IVP Books, 2019).
39. Nicholas Carr, *The Shallows: What the Internet Is Doing to Our Brains* (New York: Norton, 2010).
40. "Millennials Stand Out for Their Technology Use, but Older Generations also Embrace Digital Life," Pew Research Center, updated May 2, 2018, https://www.pewresearch.org/fact-tank/2018/05/02/millennials-stand-out-for-their-technology-use-but-older-generations-also-embrace-digital-life/.
41. "We Need to Reduce Our Dependence on Technology If We Want to Keep Innovating," Next Web, updated July 25, 2018, https://thenextweb.com/contributors/2018/07/25/we-need-to-reduce-our-dependence-on-technology-if-we-want-to-keep-innovating/; "How to Not Be a Slave to Technology," Huffington Post, updated July 28, 2013; "The Internet Has Become the External Hard Drive for Our Memories," Scientific America, updated December 1, 2013, https://www.scientificamerican.com/article/the-internet-has-become-the-external-hard-drive-for-our-memories/.
42. Damon Krukowski, *Ways of Hearing* (Cambridge, MA: MIT Press, 2019), 43, 62.

Chapter 8

1. William Brent Seales et al., "From Damage to Discovery via Virtual Unwrapping: Reading the Scroll from En-Gedi," *Science Advances* 2, no. 9 (2016), https://doi.org/10.1126/sciadv.1601247, http://advances.sciencemag.org/content/2/9/e1601247.abstract.

Appendix

1. Bob Maehre and Kenneth R. Wade, "Computer Concordances," *Ministry*, November 1987, http://adventistdigitallibrary.org/islandora/object/adl:376011/datastream/PDF/view; "The History of Bible Software Infographic," updated September 27, 2016, http://j.hn/the-history-of-bible-software-infographic/.
2. "Bible Software FAQ," updated January 1, 1998, http://bibelarbeit.net/softfaq2.htm; "Shareware: Study Tools," updated October 6, 1999, http://web.archive.org/web/19991006083257/http://www.christianshareware.net/StudyTools/.

Works Cited

"3 Takeaways from New Research on Americans and the Bible." Gospel Coalition, updated April 24, 2018. https://www.thegospelcoalition.org/blogs/trevin-wax/3-takeaways-new-research-americans-bible/.

"25th Anniversary for Bible Software Is Celebrated with New Release." *Business Wire*, updated October 11, 2005. https://www.businesswire.com/news/home/20051011005141/en/25th-Anniversary-Bible-Software-Celebrated-New-Release.

"40 under 40 Alumni." *Puget Sound Business Journal*, 2015, accessed June 10, 2016. http://www.bizjournals.com/seattle/special/2015/40under40seattle/Alumni.

"The 100 Most Read Bible Verses at Bible Gateway." *Bible Gateway*, updated May 15, 2009. https://www.biblegateway.com/blog/2009/05/the-100-most-read-bible-verses-at-biblegatewaycom/.

"About Bible Gateway." N.d., accessed February 22, 2017. https://www.biblegateway.com/about/.

"About Libronix." 2006, accessed March 2, 2017. http://web.archive.org/web/20060101130140/http://www.libronix.com/page.aspx?id=about.

"About Us." N.d., accessed February 20, 2017. https://bible.org/book/about-bibleorg.

"About Us." N.d., accessed June 1, 2019. http://theophilos.com/aboutus.htm.

"According to Mark Iv." *Time*, August 9, 1954, 68–69. http://content.time.com/time/magazine/article/0,9171,936302,00.html.

Ackerman, Rakefet, and Morris Goldsmith. "Metacognitive Regulation of Text Learning: On Screen versus on Paper." *Journal of Experimental Psychology: Applied* 17, no. 1 (2011): 18–32. http://psycnet.apa.org/journals/xap/17/1/18/.

Ackerman, Rakefet, and Tirza Lauterman. "Taking Reading Comprehension Exams on Screen or on Paper? A Metacognitive Analysis of Learning Texts under Time Pressure." *Computers in Human Behavior* 28, no. 5 (2012): 1816–28. https://doi.org/10.1016/j.chb.2012.04.023.

Ammerman, Nancy. *Congregation & Community*. New Brunswick, NJ: Rutgers University Press, 1997.

Ammerman, Nancy. "Operationalizing Evangelicalism: An Amendment." *Sociological Analysis* 43, no. 2 (1982): 170–72.

Apple. *Itunes Charts*. 2018, accessed October 15, 2018. https://www.apple.com/itunes/charts/free-apps/.

Baccino, T. *Lecture électronique: De la vision à la compréhension*. Grenoble: Presses Universitaires de Grenoble, 2004.

"Back to That Old Time Religion." *Time*, December 26, 1977, 52–58. http://content.time.com/time/subscriber/article/0,33009,919227-2,00.html.

Bailey, Sarah Pulliam. "Dozens of Evangelical Leaders Meet to Discuss How Trump Era Has Unleashed 'Grotesque Caricature' of Their Faith." Acts of Faith, *Washington Post*, April 16, 2018. https://www.washingtonpost.com/news/acts-of-faith/wp/2018/04/12/when-you-google-evangelicals-you-get-trump-high-profile-evangelicals-will-meet-privately-to-discuss-their-future/.

Baldacchino, Jean-Paul. "Markets of Piety and Pious Markets: The Protestant Ethic and the Spirit of Korean Capitalism." *Social Compass* 59, no. 3 (2012): 367–85. https://doi.org/10.1177/0037768612449721.

Balmer, Randall Herbert. *Blessed Assurance: A History of Evangelicalism in America.* Boston: Beacon Press, 1999.

Balmer, Randall Herbert. "The Wireless Gospel: Sixty-Two Years Ago, Back to the Bible Joined the Radio Revolution; Now It Is Finding New Media for Its Old Message. A Case Study in Evangelicals' Love Affair with Communications Technology." *Christianity Today* 45, no. 3 (2001): 48–53.

Barbour, Ian G. *Ethics in an Age of Technology: The Gifford Lectures, 1989–1991.* Vol. 2. San Francisco: HarperCollins, 1993.

Barna. *State of the Bible 2018: Top Findings.* July 10, 2018. https://www.barna.com/research/state-of-the-bible-2018-seven-top-findings/.

Barna. *The State of the Bible 2019.* American Bible Society, March 2019. https://1s712.americanbible.org/state-of-the-bible/stateofthebible/State_of_the_bible-2019.pdf.

Barna Group. *The Bible in America: The Changing Landscape of Perceptions and Engagement.* Ventura, CA: Barna Group, 2016.

Barna Group. *The State of the Bible 2016.* American Bible Society, 2016, accessed May 23, 2022. http://www.americanbible.org/uploads/content/State_of_the_Bible_2016_report_Politics.pdf.

Baron, Naomi S. *Words Onscreen: The Fate of Reading in a Digital World.* New York: Oxford University Press, 2015.

Barzilai-Nahon, Karine, and Gad Barzilai. "Cultured Technology: The Internet and Religious Fundamentalism." *Information Society* 21, no. 1 (2005): 25–40. https://doi.org/10.1080/01972240590895892.

Bauer, Julia. "Zondervan Acquires Religious Site Biblegateway.Com." *Grand Rapids Press*, October 28, 2008. http://www.mlive.com/grpress/business/index.ssf/2008/10/zondervan_acquires_religious_s.html.

Beaudoin, Tom. *Virtual Faith: The Irreverent Spiritual Quest of Generation X.* San Francisco: Jossey-Bass, 1998.

Bebbington, David W. *Evangelicalism in Modern Britain: A History from the 1730s to the 1980s* London: Unwin Hyman, 1989.

Benedetto, S., V. Drai-Zerbib, M. Pedrotti, G. Tissier, and T. Baccino. "E-readers and Visual Fatigue." *PLoS One* 8, no. 12 (2013): 1–7. https://doi.org/doi:10.1371/journal.pone.0083676.

Benesh, Sean. *Intrepid: Navigating the Intersection of Church Planting + Social Entrepreneurship.* N.p.: Missional Challenge, 2018.

Berger, Peter L. "Max Weber Is Alive and Well, and Living in Guatemala: The Protestant Ethic Today." *Review of Faith & International Affairs* 8, no. 4 (2010): 3–9.

Bialecki, Jon. "Between Stewardship and Sacrifice: Agency and Economy in a Southern California Charismatic Church." Entre gestion et sacrifice 14, no. 2 (2008): 372–90. https://doi.org/10.1111/j.1467-9655.2008.00507.x.

Bibb, Bryan. "Readers and Their E-Bibles: The Shape and Authority of the Hypertext Canon." In *The Bible in American Life*, edited by Philip Goff, Arthur E. Farnsley II, and Peter J. Thuesen, 266–74. Oxford: Oxford University Press, 2017.

"Bible Gateway Homepage." 1998. http://web.archive.org/web/19980110131629/http://bible.gospelcom.net/.

"Bible Labor of Years Is Done in 400 Hours." *Life*, February 18, 1957, 92.

"Bible Software FAQ." Updated January 1, 1998. http://bibelarbeit.net/softfaq2.htm.

"Bibles." 1998. http://web.archive.org/web/19980110133428/http://bible.com/bibles.html.

"Biblical Studies Foundation Homepage." 1996. http://web.archive.org/web/19961030041731/http://www.bible.org/.

Bielo, James S. "Cultivating Intimacy: Interactive Frames for Evangelical Bible Study." *Fieldwork in Religion* 3, no. 1 (2008): 51–69.

Bielo, James S. "'The Mind of Christ': Financial Success, Born-Again Personhood, and the Anthropology of Christianity." *Ethnos: Journal of Anthropology* 72, no. 3 (2007): 316–38. https://doi.org/10.1080/00141840701576935.

Bielo, James S. "On the Failure of 'Meaning': Bible Reading in the Anthropology of Christianity." *Culture and Religion* 9, no. 1 (2008): 1–21.

Bielo, James S. "Textual Ideology, Textual Practice: Evangelical Bible Reading in Group Study." In *The Social Life of Scriptures; Cross-Cultural Perspectives on Biblicism*, edited by James S. Bielo, 157–75. New Brunswick, NJ: Rutgers, 2009.

Bielo, James S. *Words upon the Word: An Ethnography of Evangelical Group Bible Study*. New York: New York University Press, 2009.

Bijker, Wiebe E. *Of Bicycles, Bakelites, and Bulbs*. Cambridge, MA: MIT Press, 1995.

Bijker, Wiebe E., Thomas P. Hughes, and Trevor Pinch. "Introduction." In *The Social Construction of Technological Systems: New Directions in the Sociology and History of Technology*, edited by Wiebe E. Bijker, Thomas P. Hughes, and Trevor Pinch, 1–10. Cambridge, MA: MIT Press, 1987.

"Billionaire Tech Mogul Bill Gates Reveals He Banned His Children from Mobile Phones until They Turned 14." *Mirror*, updated April 20, 2017. http://www.mirror.co.uk/tech/billionaire-tech-mogul-bill-gates-10265298.

Bilton, Nick. "Steve Jobs Was a Low-Tech Parent." *New York Times*, September 10, 2014. https://www.nytimes.com/2014/09/11/fashion/steve-jobs-apple-was-a-low-tech-parent.html.

"Biography of Francis Andersen." N.d., accessed September 28, 2016. http://www.aiarch.org.au/fellows.

"Blacks More Likely Than Others in U.S. to Read the Bible Regularly, See It as God's Word." Pew Research, updated May 7, 2018. http://www.pewresearch.org/fact-tank/2018/05/07/blacks-more-likely-than-others-in-u-s-to-read-the-bible-regularly-see-it-as-gods-word/.

Boone, Kathleen C. *The Bible Tells Them So: The Discourse of Protestant Fundamentalism*. Albany: State University of New York Press, 1989.

Bosker, Bianca. "The Binge Breaker." *The Atlantic*, November 2016. https://www.theatlantic.com/magazine/archive/2016/11/the-binge-breaker/501122/.

Bowles, Nellie. "The Digital Gap between Rich and Poor Kids Is Not What We Expected." *New York Times*, October 26, 2018. https://www.nytimes.com/2018/10/26/style/digital-divide-screens-schools.html.

Bracey, Glenn E., II, and Wendy Leo Moore. "Race Tests: Racial Boundary Maintenance in White Evangelical Churches." *Sociological Inquiry* 87, no. 2 (May 2017): 282–302. http://onlinelibrary.wiley.com/wol1/doi/10.1111/soin.12174/full.

Brasher, Brenda. *Give Me That Online Religion*. San Francisco: Jossey-Bass, 2001.

Brekus, Catherine A. *Sarah Osborn's World: The Rise of Evangelical Christianity in Early America*. New Haven, CT: Yale University Press, 2013.

Brenner, Philip S. "Cross-National Trends in Religious Service Attendance." *Public Opinion Quarterly* 80, no. 2 (Summer 2016): 563–83. https://doi.org/10.1093/poq/nfw016.

Brody, David, and Scott Lamb. *The Faith of Donald J. Trump: A Spiritual Biography.* New York: Broadside Books, 2018.
Brooke, Tal. *Virtual Gods: The Seduction of Power and Pleasure in Cyberspace*. Eugene, OR: Harvest House, 1997.
Brown, Callum G. *The Death of Christian Britain: Understanding Secularisation, 1800–2000.* London: Routledge, 2001.
Brown, Charles. "Selling Faith: Marketing Christian Popular Culture to Christian and Non-Christian Audiences." *Journal of Religion and Popular Culture* 24, no. 1 (Spring 2012). https://doi.org/10.3138/jrpc.24.1.113.
Bruce, Steve. "Identifying Conservative Protestantism." *Sociological Analysis* 44, no. 1 (1983): 65–70.
Brumley, Jeff. "Name Changes Challenge Churches on Baptist Identity." *Baptist Global News*, October 7, 2014. https://baptistnews.com/ministry/congregations/item/29312-name-changes-challenge-churches-on-baptist-identity.
Burch, Jack. "The Use of a Computer in New Testament Text Criticism." *Restoration Quarterly* 8, no. 2 (1965): 119–25.
Bureau of Labor Statistics. *American Time Use Survey—2014 Results.* US Department of Labor, June 24, 2015. https://www.bls.gov/news.release/archives/atus_06242015.pdf.
Burleigh, Nina. "Does God Believe in Trump? White Evangelicals Are Sticking with Their 'Prince of Lies.'" *Newsweek*, October 5, 2017. http://www.newsweek.com/2017/10/13/donald-trump-white-evangelicals-support-god-677587.html.
Burton, Delores M. "Automated Concordances and Word Indexes: The Fifties." *Computers and the Humanities* 1, no. 15 (1981): 1–14.
Busch, Laura. "To Come to a Correct Understanding of Buddhism: A Case Study on Spiritualizing Technology, Religious Authority, and the Boundaries of Orthodoxy and Identity in a Buddhist Web Forum." *New Media & Society* 13, no. 1 (2011): 58–74. https://doi.org/10.1177/1461444810363909.
Byers, Andrew. *Theomedia: The Media of God and the Digital Age*. Havertown: Lutterworth Press, 2014.
Cadwalladr, Carole, and Emma Graham-Harrison. "Revealed: 50 Million Facebook Profiles Harvested for Cambridge Analytica in Major Data Breach." *The Guardian*, March 17, 2018. https://www.theguardian.com/news/2018/mar/17/cambridge-analytica-facebook-influence-us-election.
Campbell, Heidi A. *Digital Creatives and the Rethinking of Religious Authority Online.* London: Routledge, 2021.
Campbell, Heidi A. *Exploring Religious Community Online: We Are One in the Network.* New York: Peter Lang, 2005.
Campbell, Heidi A. "Religious Communication and Technology." *Annals of the International Communication Association* 41, nos. 3–4 (2017): 1–7. https://doi.org/10.1080/23808985.2017.1374200.
Campbell, Heidi A. "Surveying Theoretical Approaches within Digital Religion Studies." *New Media & Society* 19, no. 1 (2016): 15–24. https://doi.org/10.1177/1461444816649912.
Campbell, Heidi A. "Understanding the Relationship between Religion Online and Offline in a Networked Society." *Journal of the American Academy of Religion* 80, no. 1 (2012): 64–93.
Campbell, Heidi A. *When Religion Meets New Media*. New York: Routledge, 2010.
Campbell, Heidi A. "Who's Got the Power? Religious Authority and the Internet." *Journal of Computer-Mediated Communication* (2007): 1043–62.

Campbell, Heidi A., and Brian Altenhofen. "Digitizing Research in the Sociology of Religion." In *Digital Methodologies in the Sociology of Religion*, edited by Sariya Cheruvallil-Contractor and Suda Shakkour, 1–12. New York: Bloomsbury, 2016.

Campbell, Heidi A., Brian Altenhofen, Wendi Bellar, and Kyong James Cho. "There's a Religious App for That! A Framework for Studying Religious Mobile Applications." *Mobile Media & Communication* 2, no. 2 (2014): 154–72.

Campbell, Heidi A., and Kyong James Cho. "Religious Use of Mobile Phones." In *Encyclopedia of Mobile Phone Behavior*, edited by Zheng Yan, 308–21. Hersey, PA: Information Science Reference, 2015.

Campbell, Heidi A., and John Dyer, eds. *Ecclesiology for a Digital Church*. London: SCM Press, 2022.

Campbell, Heidi A., and Oren Golan. "Creating Digital Enclaves: Negotiation of the Internet among Bounded Religious Communities." *Media, Culture & Society* 33, no. 5 (2011): 709–24. https://doi.org/10.1177/0163443711404464.

Campbell, Heidi A., and Mia Lövheim. "Rethinking the Online-Offline Connection in the Study of Religion Online." *Information, Communication & Society* 14, no. 8 (2011): 1083–96.

Cao, Nanlai. "Boss Christians: The Business of Religion in the 'Wenzhou Model' of Christian Revival." *China Journal* 59 (2008): 63–87. https://doi.org/10.2307/20066380. http://www.jstor.org/stable/20066380.

Capps, Walter H. *The New Religious Right: Piety, Patriotism, and Politics*. Columbia: University of South Carolina Press, 1990.

Carr, Nicholas. "Is Google Making Us Stupid?" *The Atlantic*, July/August 2008. https://www.theatlantic.com/magazine/archive/2008/07/is-google-making-us-stupid/306868/.

Carr, Nicholas. *The Shallows: What the Internet Is Doing to Our Brains*. New York: Norton, 2010.

"Celebrating an Online Bible Legacy: Bible Gateway." N.d., accessed June 12, 2018. https://www.biblegateway.com/article.

Challies, Tim. "Logos Bible Software 3." *Journal of Modern Ministry* 3, no. 3 (Fall 2006). https://www.logos.com/press/reviews/challies.

Challies, Tim. *The Next Story: Life and Faith after the Digital Explosion*. Grand Rapids, MI: Zondervan, 2011.

Chen, Guang, Wei Cheng, Ting-Wen Chang, Xiaoxia Zheng, and Ronghuai Huang. "A Comparison of Reading Comprehension across Paper, Computer Screens, and Tablets: Does Tablet Familiarity Matter?" *Journal of Computers in Education* 1 (2014): 213–25. https://doi.org/10.1007/s40692-014-0012-z.

Chow, Alexander. "What Has Jerusalem to Do with the Internet? World Christianity and Digital Culture." *International Bulletin of Mission Research* (2021).

Christerson, Brad, and Richard Flory. *The Rise of Network Christianity: How Independent Leaders Are Changing the Religious Landscape*. Oxford: Oxford University Press, 2017.

"Church and Technology: A Survey of Ontario Churches." *Tyndale: The Magazine*, Fall 2011.

"Church Planting 1: The Movement." Startup from Gimlet Media, updated July 6, 2018. https://www.gimletmedia.com/startup/church-planting-1-the-movement.

Clark, C. *Redefining or Undermining? The Role of Technology in the Reading Lives of Children and Young People: Findings from the National Literacy Trust's Annual Survey 2012*. London: National Literacy Trust, 2012.

"Clinton Maintains Double-Digit (51% vs. 36%) Lead over Trump." Public Religion Research Institute / Brookings Survey, updated October 19, 2016. https://www.prri.org/research/prri-brookings-oct-19-poll-politics-election-clinton-double-digit-lead-trump/.

Cockburn, Cynthia, and Susan Omrud. *Gender and Technology in the Making.* London: Sage, 1993.

"[comp.infosystems.www] Anyone Want to Test a Gateway?" Updated December 28, 1993. https://groups.google.com/forum/#!msg/comp.archives/qDBF4nb7jSk/AxgqJK6BsusJ.

"Computer Software Can Aid Preachers." *Preaching*, 1987, accessed July 1, 2016. http://beta.preaching.com/resources/articles/11566947/.

Connolly, William E. *Capitalism and Christianity, American Style.* Durham, NC: Duke University Press Books, 2008.

Connolly, William E. "The Evangelical-Capitalist Resonance Machine." *Political Theory* 33, no. 6 (2005): 869–86. http://www.jstor.org/stable/30038467.

Conyers, A. J. "Three Sources of the Secular Mind." *Journal of the Evangelical Theological Society* 41, no. 2 (June 1, 1998): 313–21.

Corbin, Wyndy. "The Impact of the American Dream on Evangelical Ethics." *Cross Currents* 55, no. 3 (Fall 2005): 340–50.

Cowan, Douglas. *Cyberhenge: Modern Pagans on the Internet.* New York: Routledge, 2004.

Cox, Daniel, and Robert P. Jones. *America's Changing Religious Identity.* Public Religion Research Institute, September 6, 2017. https://www.prri.org/research/american-religious-landscape-christian-religiously-unaffiliated/.

Crapanzano, Vincent. *Serving the Word: Literalism in America from the Pulpit to the Bench.* New York: New Press, 2000.

Crossley, Samuel. "Recent Developments in the Definition of Evangelicalism." *Foundations (Affinity)* 70 (2016): 112–33.

Crouch, Andy. *Culture Making: Recovering Our Creative Calling.* Downers Grove, IL: IVP Books, 2013.

Culkin, John. "Each Culture Develops Its Own Sense Ratio to Meet the Demands of Its Environment." In *Mcluhan Hot and Cool: A Primer for the Understanding of Mcluhan*, edited by Gerald Emanuel Stearn, 49–57. New York: New American Library, 1967.

Cushman, W. H. "Reading from Microfiche, Vdt and the Printed Page: Subjective Fatigue and Performance." *Human Factors* 28, no. 1 (1986): 63–73.

Daughters, Kenneth Alan. "Logos Bible Software 2.0 Level 4." *Emmaus Journal* 4, no. 1 (1995): 92–93.

Daughters, Kenneth Alan. "Review of Computer Concordance Software." *Emmaus Journal* 2, no. 1 (1993): 77–93.

Davenport, Thomas H., and John C. Beck. *The Attention Economy: Understanding the New Currency of Business.* Boston: Harvard Business School Press, 2001.

Davids, Peter H. *The Letters of 2 Peter and Jude.* Pillar. Grand Rapids, MI: Eerdmans, 2006.

Davie, Grace. *The Sociology of Religion.* 2nd ed. London: Sage, 2013.

Dawson, Lorne L. "Religion and the Quest for Virtual Community." In *Religion Online: Finding Faith on the Internet*, edited by Lorne L. Dawson and Douglas E. Cowan, 75–92. New York: Routledge, 2005.

Dawson, Lorne L. "Researching Religion in Cyberspace: Issues and Strategies." In *Religion on the Internet: Research Prospects and Promises*, edited by Douglas E. Cowan and Jeffrey K. Hadden, 24–54. New York: HAI Press, 2000.

Day, Abby. *Believing in Belonging: Belief and Social Identity in the Modern World.* Oxford: Oxford University Press, 2011.

Day, Elizabeth. "How Selfies Became a Global Phenomenon." *The Guardian*, July 13, 2013. http://www.theguardian.com/technology/2013/jul/14/how-selfies-became-a-global-phenomenon.

"Dear Pastor, Bring Your Bible to Church." The Gospel Coalition, August 19 , 2013. http://thegospelcoalition.org/article/dear-pastor-bring-your-bible-to-church.

Deegan, Marilyn, and Kathryn Sutherland. *Transferred Illusions: Digital Technology and the Forms of Print.* New York: Routledge, 2009.

Dehaene, Stanislas. *Reading in the Brain: The Science and Evolution of a Human Invention.* New York: Viking, 2009.

DeStefano, Diana, and Jo-Anne LeFevre. "Cognitive Load in Hypertext Reading: A Review." *Computers in Human Behavior* 23, no. 3 (2007): 1616–41.

DeYmaz, Mark. *Building a Healthy Multi-ethnic Church: Mandate, Commitments, and Practices of a Diverse Congregation.* San Francisco: Jossey-Bass / John Wiley, 2007.

Deyo, Steve. "Cyber-boost Your Faith: The Latest Bible Study Software Will Empower Your Study at Any Level." *Christianity Today*, April 27, 1998, S15.

Dillon, Andrew. "Reading from Paper versus Screens: A Critical Review of the Empirical Literature." *Ergonomics* 35, no. 10 (1992): 1297–326. https://www.ischool.utexas.edu/~adillon/Journals/Reading.htm.

Dinwoodie, C. "Christianity in the Computer Age." *Scottish Journal of Theology* 18, no. 2 (1965): 204–18.

Dochuk, Darren. *From Bible Belt to Sunbelt: Plain-Folk Religion, Grassroots Politics, and the Rise of Evangelical Conservatism.* New York: Norton, 2010.

Dougherty, Kevin D., Jenna Griebel, Mitchell J. Neubert, and Jerry Z. Park. "A Religious Profile of American Entrepreneurs." *Journal for the Scientific Study of Religion* 52, no. 2 (2013): 401–9. https://doi.org/10.1111/jssr.12026.

Dubis, M. "The Bible in Your Palm: Biblical Studies Software for Palm Os Handheld Devices." *Journal of Religious & Theological Information* 4, no. 4 (2001): 3–10.

Dudley, Jonathan. *Broken Words: The Abuse of Science and Faith in American Politics.* New York: Crown Publishers, 2011.

Duduit, Michael. "Using Your Computer in Sermon Preparation." *Preaching*, September 1, 1990. http://www.preaching.com/resources/articles/11567226/.

Du Mez, Kristin Kobes. *Jesus and John Wayne: How White Evangelicals Corrupted a Faith and Fractured a Nation.* New York: Liveright, 2020.

Dye, T. Wayne. "The Eight Conditions of Scripture Engagment: Social and Cultural Factors Necessary for Vernacular Bible Translation to Achieve Maximum Effect." *International Journal of Frontier Missiology* 26, no. 2 (2009): 89–98.

Dye, T. Wayne. "Scripture in an Accessible Form: The Most Common Avenue to Increased Scripture Engagement." *International Journal of Frontier Missiology* 26, no. 3 (2009): 123–28.

Dyer, John. *From the Garden to the City: The Redeeming and Corrupting Power of Technology.* Grand Rapids, MI: Kregel, 2011.

Dyer, John. "The Habits and Hermeneutics of Digital Bible Readers: Comparing Print and Screen Engagement, Comprehension, and Behavior." *Journal of Religion, Media and Digital Culture* 8, no. 2 (2019): 181–205. https://doi.org/https://doi.org/10.1163/21659214-00802001.

Earley, Justin Whitmel. *The Common Rule: Habits of Purpose for an Age of Distraction.* Downers Grove, IL: IVP Books, 2019.

"Ease of Use, Speed Linked in New Lifeway Bible Software." Baptist Press, updated October 17, 2003. http://www.bpnews.net/16884/ease-of-use-speed-linked-in-new-lifeway-bible-software.

Edge, David. "The Social Shaping of Technology." Edinburgh PICT Working Paper no. 1, Edinburgh University, 1988.

Edsall, Thomas B. "Trump Says Jump. His Supporters Ask, How High?" *New York Times*, September 14, 2017. https://www.nytimes.com/2017/09/14/opinion/trump-republicans.html.

Edwards, John. "The Disk-Based Bible." *Popular Computing*, 12, no. 2, 1983. https://archive.org/stream/militarychaplain35unse/militarychaplain35unse_djvu.txt.

Eisenstein, Elizabeth L. *The Printing Press as an Agent of Change: Communications and Cultural Transformations in Early Modern Europe*.2 vols. Cambridge: Cambridge University Press, 1979.

Eklundh, Kerstin Severinson. "Problems in Achieving a Global Perspective of the Text in Computer-Based Writing." *Instructional Science* 21, no. 1 (1992): 73–84.

Ellis, Yvonne, Bobbie Daniels, and Andres Jauregui. "The Effect of Multitasking on the Grade Performance of Business Students." *Research in Higher Education Journal* 8, no. 1 (2010): 1–10.

Ellison, John. "Review of Christianity in the Computer Age." *Journal of Biblical Literature* 84, no. 4 (1965): 190–91.

Ellul, Jacques. *Propaganda: The Formation of Men's Attitudes*. New York: Vintage, 1972.

Ellul, Jacques. *The Technological Society*. New York: Vintage, 1964.

Emerson, Michael O, and Karen Chai Kim. "Multiracial Congregations: An Analysis of Their Development and a Typology." *Journal for the Scientific Study of Religion* 42, no. 2 (2003): 217–27.

Ericsson. *Mobility Report*, June 3, 2018. https://www.ericsson.com/en/mobility-report.

Estes, Douglas. *Simchurch: Being the Church in the Virtual World*. Grand Rapids, MI: Zondervan, 2009.

Etherington, Norman. *Missions and Empire*. The Oxford History of the British Empire Companion Series. Oxford: Oxford University Press, 2005.

"Exposing America's Biggest Hypocrites: Evangelical Christians." *Huffington Post*, updated November 24, 2017. https://www.huffingtonpost.com/entry/exposing-americas-biggest-hypocrites-evangelical_us_5a184f0ee4b068a3ca6df7ad.

Eyal, Nir. *Hooked: How to Build Habit-Forming Products*. New York: Penguin, 2014.

Eyland, E. Ann. "Revelations from Word Counts." In *Perspectives on Language and Text: Essays and Poems in Honor of Francis I. Andersen's Sixtieth Birthday, July 28, 1985*, edited by Edward G. Newing and Edgar W Conrad. Winona Lake, IN: Eisenbrauns, 1987.

Fackler, Mark. "The Second Coming of Holy Writ: Niche Bibles and the Manufacture of Market Segments." In *New Paradigms for Bible Study: The Bible in the Third Millennium*, edited by Robert M. Fowler, Edith Blumhofer, and Fernando F. Segovia, 71–88. New York: T & T Clark, 2004.

Fackre, Gabriel. *The Religious Right and Christian Faith*. Grand Rapids, MI: Eerdmans, 1982.

"Faithlife CEO Bob Pritchett on Balancing the Principles of Faith, Science and Entrepreneurship." *Beliefnet*, 2015, accessed February 10, 2017. https://www.beliefnet.com/columnists/faithmediaandculture/2015/09/faithlife-ceo-bob-pritchett-on-balancing-the-principles-of-faith-science-and-entrepreneurship.html.

"Faithlife's Bob Pritchett Ranked among Top 25 CEOs in 2015." *Logos*, updated June 29, 2015. https://blog.faithlife.com/blog/2015/06/faithlifes-bob-pritchett-among-top-25-ceos-of-2015/.

Feenberg, Andrew. "What Is Philosophy of Technology?" Lecture for the Komaba undergraduates, 2003, accessed October 15, 2015. https://www.sfu.ca/~andrewf/books/What_is_Philosophy_of_Technology.pdf.

Fekete, John. "McLuhanacy: Counterrevolution in Cultural Theory." *Telos* no. 15 (Spring 1973): 75–123.

Ferguson, Kathy E. "Bringing Gender into the Evangelical-Capitalist Resonance Machine." *Political Theology* 12, no. 2 (2011): 184–94. https://doi.org/10.1558/poth.v12i2.184.

"The First-Ever Hashtag, @-Reply and Retweet, as Twitter Users Invented Them." *Quartz*, updated October 15, 2013. https://qz.com/135149/the-first-ever-hashtag-reply-and-retweet-as-twitter-users-invented-them/.

Fisher, Linford D. "Evangelicals and Unevangelicals: The Contested History of a Word, 1500–1950." *Religion and American Culture* 26, no. 2 (Summer 2016): 184–226. https://doi.org/10.1525/rac.2016.26.2.184.

Flanigan, Abraham, and Markeya Peteranetz. "Digital Distraction across Courses: Self-Regulation of Digital Device Use in Favorite versus Least Favorite Courses." Paper presented at Annual Meeting of the American Educational Research Association, Toronto, April 5–9, 2019.

Flory, Richard W., and Donald E. Miller. *Finding Faith: The Spiritual Quest of the Post-boomer Generation*. New Brunswick, NJ: Rutgers University Press, 2008.

Fong, Bruce W. *Racial Equality in the Church: A Critique of the Homogeneous Unit Principle in Light of a Practical Theology Perspective*. Lanham, MD: University Press of America, 1996.

Ford, David G., Joshua L. Mann, and Peter M. Phillips. *The Bible and Digital Millennials*. New York: Routledge, 2019.

Friesen, Amanda. "How American Men and Women Read the Bible." In *The Bible in American Life*, edited by Philip Goff, Arthur E. Farnsley II, and Peter J. Thuesen, 266–74. Oxford: Oxford University Press, 2017.

Fulk, Janet. "Social Construction of Communication Technology." *Academy of Management Journal* 36, no. 5 (1993): 921–50.

Fulks, Jeffery, Randy Petersen, and John Farquhar Plake. *The State of the Bible 2022*. American Bible Society, March 2022. https://1s712.americanbible.org/state-of-the-bible/stateofthebible/State_of_the_bible-2022.pdf.

"Futures for the Bible Gateway." 1997, accessed December 10, 2017. http://web.archive.org/web/19971210071327/http://bible.gospelcom.net/bg/futures.html.

Gabriel, Trip. "Donald Trump, Despite Impieties, Wins Hearts of Evangelical Voters." *New York Times*, February 28, 2016. https://www.nytimes.com/2016/02/28/us/politics/donald-trump-despite-impieties-wins-hearts-of-evangelical-voters.html.

Gamble, Harry Y. *Books and Readers in the Early Church: A History of Early Christian Texts*. New Haven, CT: Yale University Press, 1995.

Gardner, Howard, and Katie Davis. *The App Generation: How Today's Youth Navigate Identity, Intimacy, and Imagination in a Digital World*. New Haven, CT: Yale University Press, 2013. http://books.google.com/books?isbn=9780300196214.

Garrett, Lynn. "Harper Acquires Bible Software Company." *Publishers Weekly*, May 6, 2014. http://www.publishersweekly.com/pw/by-topic/industry-news/religion/article/62160-harper-acquires-bible-software-company.html.

Gay, Craig M. *Modern Technology and the Human Future: A Christian Appraisal*. Downers Grove, IL: InterVarsity Press, 2018.

Gbadamosi, Ayantunji. "Exploring the Growing Link of Ethnic Entrepreneurship, Markets, and Pentecostalism in London (UK): An Empirical Study." *Society and Business Review* 10 (2015): 150–69. https://doi.org/10.1108/SBR-11-2014-0053.

Gee, Buck, and Denise Peck. *The Illusion of Asian Success: Scant Progress for Minorities in Cracking the Glass Ceiling from 2007–2015*. Ascend Foundation Research, 2017, accessed May 18, 2002. https://c.ymcdn.com/sites/www.ascendleadership.org/resource/resmgr/research/TheIllusionofAsianSuccess.pdf.

Gloege, Timothy E. W. *Guaranteed Pure: The Moody Bible Institute, Business, and the Making of Modern Evangelicalism*. Chapel Hill: University of North Carolina Press, 2015.

Gnatek, Tim. "A New Bible, Palmtop Version, Can Keep Track of Studies." *New York Times*, December 2, 2004. http://www.nytimes.com/2004/12/02/technology/circuits/a-new-bible-palmtop-version-can-keep-track-of-studies.html.

Goff, Philip, Arthur E. Farnsley II, and Peter J. Thuesen. "The Bible in American Life Today." In *The Bible in American Life*, edited by Philip Goff, Arthur E. Farnsley II, and Peter J. Thuesen, 5–34. Oxford: Oxford University Press, 2017.

Gonzalez, Heather. "NAE's Beginning." *Evangelicals* 3, no. 3 (Winter 2017): 20–21.

Goodwin, Robert. "Christianity in the Computer Age." *Drew Gateway* 39, no. 3 (1969): 153–57.

"Google's Gmail Controversy Is Everything People Hate About Silicon Valley." Updated July 3, 2018, accessed November 12, 2019. https://www.cnet.com/news/privacy/googles-gmail-controversy-is-everything-wrong-with-silicon-valley/.

Gordon, T. David. *Why Johnny Can't Preach: The Media Have Shaped the Messengers*. Phillipsburg, NJ: P & R Pub., 2009.

Gordon, T. David. *Why Johnny Can't Sing Hymns: How Pop Culture Rewrote the Hymnal*. Phillipsburg, NJ: P & R Pub., 2010.

Graham, Billy. *Just as I Am: The Autobiography of Billy Graham*. 10th anniversary ed. New York: HarperOne, 2007.

Graham, Roderick, and Kyungsub Stephen Choi. "Explaining African-American Cell Phone Usage through the Social Shaping of Technology Approach." *Journal of African American Studies* 20 (2016): 19–34.

Green, Robert W. *Protestantism and Capitalism: The Weber Thesis and Its Critics*. Boston: Heath, 1959.

Gribben, Crawford. "Puritan Subjectivities: The Conversion Debate in Cromwellian Dublin." In *Converts and Conversion in Ireland, 1650–1850*, edited by Michael L. Brown, McGrath Carlies Ivar, and Tom P. Power, 79–106. Dublin: Four Courts Press, 2005.

Griswold, Eliza. "Millennial Evangelicals Diverge from Their Parents' Beliefs." *New Yorker*, August 27, 2018. https://www.newyorker.com/news/on-religion/millennial-evangelicals-diverge-from-their-parents-beliefs.

Guerrant, William C. *Organic Wesley: A Christian Perspective on Food, Farming, and Faith*. Franklin, TN: Seedbed Publishing, 2015.

Guest, Mathew. *Evangelical Identity and Contemporary Culture: A Congregational Study in Innovation*. Milton Keynes, England: Paternoster, 2007.

Gutjahr, Paul. "Protestant English-Langauge Bible Publishing and Translation." In *The Oxford Handbook of the Bible in America*, edited by Paul Gutjahr, 3–18. Oxford: Oxford University Press, 2018.

Hadden, Jeffrey K., and Douglas E. Cowan, eds. *Religion on the Internet: Research Promises and Prospects*. Bingley, UK: Emerald Group, 2000.

Hall, Josh. "Technophobic Pastors of Tech-Addicted Churches." *The Evangelicals* 4, no. 1 (Spring/Summer 2018): 18–20.

Hals, Tom. "Bible.Com Investor Sues Company for Lack of Profit." *Reuters*, October 21, 2010. http://www.reuters.com/article/us-biblecom-lawsuit-idUSTRE69K42D20101021.

"Hand Held Portable Computer in the Form of a Book." Updated September 19, 1983. https://patents.justia.com/patent/D284966.

Hannah, John D. *An Uncommon Union: Dallas Theological Seminary and American Evangelicalism*. Grand Rapids, MI: Zondervan, 2009.

Harbin, Duane. "Fiat Lux." In *Formatting the Word of God: An Exhibition at Bridwell Library*, edited by Valerie R. Hotchkiss and Charles C. Ryrie. Dallas, TX: Bridwell Library, 1998. https://web.archive.org/web/20051202042630/http://www.smu.edu/bridwell/publications/ryrie_catalog/xiii_1.htm.

Hardesty, Nancy. "Evangelical Women Face Their Homophobia." *Christian Century* 103, no. 26 (1986): 768.

Hardiman, Clayton. "Gospel Communications Online Sold." *Muskegon Chronicle*, October 29, 2008. http://blog.mlive.com/chronicle/2008/10/gospel_communications_online_s.html.

"HarperCollins Christian Publishing Acquires Olive Tree." HarperCollins, updated May 5, 2014. https://www.harpercollinschristian.com/harpercollins-christian-publishing-acquires-olive-tree/.

Hawkins, Lori. "Scriptures on Screen." *Austin American-Statesman*, December 10, 1993, C1.

Hayes, John H. "The Evangelical Ethos and the Spirit of Capitalism." *Perspectives in Religious Studies* 39, no. 3 (2012): 205–17.

Haynes, Robert Ellis. *Consuming Mission: Towards a Theology of Short-Term Mission and Pilgrimage*. Eugene, OR: Pickwick Publications, 2018.

Helland, Christopher. "Online Religion as Lived Religion: Methodological Issues in the Study of Religious Participation on the Internet." *Heidelberg Journal of Religions on the Internet* 1, no. 1 (2005): 1–16.

Helland, Christopher. "Online-Religion/Religion-Online and Virtual Communities." In *Religion on the Internet: Research Prospects and Promises*, edited by Douglas E. Cowan and Jeffrey K. Hadden, 205–23. New York: HAI Press, 2000.

Helm, Joe. "Jerry Falwell Jr. Can't Imagine Trump 'Doing Anything That's Not Good for the Country.'" *Washington Post*, January 1, 2019. https://www.washingtonpost.com/lifestyle/magazine/jerry-falwell-jr-cant-imagine-trump-doing-anything-thats-not-good-for-the-country/2018/12/21/6affc4c4-f19e-11e8-80d0-f7e1948d55f4_story.html.

Hewitt, Hugh. "Why Christians Will Stick with Trump." *Washington Post*, October 5, 2017. https://www.washingtonpost.com/opinions/why-christians-will-stick-with-trump/2017/10/05/7d7d2bb6-a922-11e7-850e-2bdd1236be5d_story.html.

Hicks, John Mark. "Numerical Growth in the Theology of Acts: The Role of Pragmatism, Reason and Rhetoric." Paper presented at the Forty-Seventh National Conference of the Evangelical Theological Society, Philadelphia, PA, November 16–18, 1995.

Hiemstra, Rick. *Confidence Conversation and Community: Bible Engagement in Canada, 2013*. Toronto: Faith Today Publications, 2014. http://files.evangelicalfellowship.ca/research/CBES-Report-Confidence-Conversation-and-Community.pdf.

Hillborn, David. "Principled Unity or Pragmatic Compromise? The Challenge of Pan-Evangelical Theology." *Evangel* 22, no. 3 (2004): 80–91.

Hillesund, Terje. "Digital Reading Spaces: How Expert Readers Handle Books, the Web and Electronic Paper." *First Monday* 15, nos. 4–5 (2010). https://doi.org/10.5210/fm.v15i4.2762.

Hindley, Meredith. "The Rise of the Machines: NEH and the Digital Humanities: The Early Years." *Humanities* 34, no. 4 (2013). http://www.neh.gov/humanities/2013/julyaugust/feature/the-rise-the-machines.

Hine, Christine. *Virtual Ethnography*. London: Sage, 2000.

Hinson, E. Glenn. "William Carey and Ecumenical Pragmatism." *Journal of Ecumenical Studies* 17, no. 2 (1980): 73–83.

Hinson, G. "Christianity in the Computer Age." *Review & Expositor* 62, no. 4 (1965): 493–94.

Hipps, Shane. *Flickering Pixels: How Technology Shapes Your Faith*. Grand Rapids, MI: Zondervan, 2009.

Hipps, Shane. *The Hidden Power of Electronic Culture: How Media Shapes Faith, the Gospel, and Church*. El Cajon, CA: Youth Specialties, 2006.

"The History of Bible Software Infographic." Updated September 27, 2016. http://j.hn/the-history-of-bible-software-infographic/.

"A History of Ellis Enterprises." 2000, accessed August 12, 2017. https://web.archive.org/web/20000919233657/http://www.biblelibrary.com/company.htm.

"The History of Swordsearcher." Updated March 2018. https://www.swordsearcher.com/history-of-swordsearcher-bible-software.html.

Højsgaard, Morten T., and Margit Warburg. "Introduction: Waves of Research." In *Religion and Cyberspace*, ed. Morten T. Højsgaard and Margit Warburg, 1–11. New York: Routledge, 2005.

Hoover, Christine. *The Church Planting Wife: Help and Hope for Her Heart*. Chicago: Moody Publishers, 2013.

Howard, Evan. "*Lectio Divina* in the Evangelical Tradition." *Journal of Spiritual Formation & Soul Care* 5, no. 1 (2012): 56–77.

"How to Build Your Own Bible." *Smile Politely*, updated August 12, 2009. http://www.smilepolitely.com/opinion/how_to_build_your_own_bible/.

"How to Not Be a Slave to Technology." *Huffington Post*, updated July 28, 2013.

Hsu, Jeffrey. *A Comprehensive Guide to Computer Bible Study: Up-to-Date Information on the Best Software and Techniques*. Dallas, TX: Word Publishing, 1993.

Huang, Travis K., and Fu Fong-Ling. "Understanding User Interface Needs of E-commerce Web Sites." *Behaviour & Information Technology* 28, no. 5 (2009): 461–69. https://doi.org/10.1080/01449290903121378.

Hubbard, David Allan. *What We Evangelicals Believe*. Pasadena, CA: Fuller Seminary Press, 1991.

Hughes, John. *Bits, Bytes and Biblical Studies*. Grand Rapids, MI: Zondervan, 1987.

Hunter, David. "How They Got the Bible on Disk." *Softalk Mag*, May 9, 1982.

Hunter, James Davison. "Operationalizing Evangelicalism: A Review, Critique & Proposal." *Sociological Analysis* 42, no. 4 (1981): 363–72.

Hutchby, Ian. "Technologies, Texts and Affordances." *Sociology* 35, no. 2 (2001): 441–56. https://doi.org/10.1017/S0038038501000219.

Hutchings, Tim. "Creating Church Online: An Ethnographic Study of Five Internet-Based Christian Communities." PhD diss., Durham University, 2010.

Hutchings, Tim. *Creating Church Online: Ritual, Community and New Media*. London: Routledge, 2017.

Hutchings, Tim. "Design and the Digital Bible: Persuasive Technology and Religious Reading." *Journal of Contemporary Religion* 32, no. 2 (2017): 205–19. https://doi.org/10.1080/13537903.2017.1298903.

Hutchings, Tim. "E-reading and the Christian Bible." *Studies in Religion / Studies Religieuses* 44, no. 4 (October 16, 2015): 423–40.

Hutchings, Tim. "Now the Bible Is an App: Digital Media and Changing Patterns of Religious Authority." In *Religion, Media, and Social Change*, edited by Kennet Granholm, Marcus Moberg, and Sofia Sjö, 143–61. New York: Routledge, 2014.

Hutchings, Tim. "'The Smartest Way to Study the Word': Protestant and Catholic Approaches to the Digital Bible." In *Negotiating Religious Visibility in Digital Media*, edited by Miriam Diez Bosch, Josep Lluís Micó, and Josep Maria Carbonell, 57–68. Barcelona: Facultat de Comunicació i Relacions Internacionals Blanquerna, 2015.

Hutchings, Tim. "Studying Apps: Research Approaches to the Digital Bible." In *Digital Methodologies in the Sociology of Religion*, edited by Sariya Cheruvallil-Contractor and Suda Shakkour, 97–108. New York: Bloomsbury, 2015.

"Introducing the Every Day Bible App." Updated December 3, 2014. https://blog.faithlife.com/blog/2014/12/introducing-the-every-day-bible-app/.

"Island Code Works Homepage." 1999, accessed February 24, 2017. http://web.archive.org/web/19991127121128/http://islandcodeworks.com/.

"J. Hampton Keathley, III, Author Page." n.d., accessed February 20, 2017. https://bible.org/users/j-hampton-keathley-iii.

Jacobs, Alan. "Why Bother with Marshall McLuhan?" *New Atlantis*, Spring 2011, 123–35.

"James R. Solakian vs. Bible.Com, Inc." Updated October 19, 2010. https://picker.typepad.com/files/bible.com.pdf.

Jefferson, Thomas, Harry R. Rubenstein, Barbara Clark Smith, and Janice Stagnitto Ellis. *The Jefferson Bible: The Life and Morals of Jesus of Nazareth, Extracted Textually from the Gospels in Greek, Latin, French & English*. Washington, DC: Smithsonian Books, 2011.

Jeffrey, David Lyle. *People of the Book: Christian Identity and Literary Culture*. Grand Rapids, MI: Eerdmans, 1996.

Johansson, Anna. "We Need to Reduce Our Dependence on Technology If We Want to Keep Innovating." *Next Web*, updated July 25, 2018. https://thenextweb.com/contributors/2018/07/25/we-need-to-reduce-our-dependence-on-technology-if-we-want-to-keep-innovating/.

Johnson, Stephen D., and Joseph B. Tamney. "The Christian Right and the 1980 Presidential Election." *Journal for the Scientific Study of Religion* 21, no. 2 (1982): 123–31. https://doi.org/10.2307/1385498.

Juzwik, Mary M. "American Evangelical Biblicism as Literate Practice: A Critical Review." *Reading Research Quarterly* 49, no. 3 (2014): 335–49. https://doi.org/10.1002/rrq.72.

Kang, Minjeong, and Juhyun Eune. "Design Framework for Multimodal Reading Experience in Cross-Platform Computing Devices—Focus on a Digital Bible." Paper presented at the DRS 2012, Chulalongkorn University, Bangkok.

Kelderman, Eric, and Dan Bauman. "'Fear, Intimidation, Bullying': Inside One of the Most Scathing Accreditation Reports in Recent Memory." *Chronicle*, December 21, 2018, A23–25.

Keller, Timothy, and Katherine Leary Alsdorf. *Every Good Endeavour: Connecting Your Work to God's Plan for the World*. London: Hodder & Stoughton, 2014.

Kellner, Mark. *God on the Internet*. Hoboken, NJ: John Wiley & Sons, 1996.

Kellstedt, Lyman A., John C. Green, James L. Guth, and Corwin E. Smidt. "Religious Voting Blocs in the 1992 Election: The Year of the Evangelical?" *Sociology of Religion* 55, no. 3 (Fall 1994): 307–26.

Kever, Jonathan. "The Year's Best Software for Preachers: Version 2003." *Preaching*, 2003, accessed August 22, 2022. https://www.preaching.com/articles/the-years-best-software-for-preachers-version-2003/.

Kivi, Lea Karen. "#Churchtoo." *America* 219, no. 12 (November 26. 2018): 28–33.

Klein, Hans, and Daniel Lee Kleinman. "The Social Construction of Technology: Structural Considerations." *Science, Technology, & Human Values* 27, no. 1 (2002): 28–52.

Klein, Robert W. "Something New under the Sun: Computer Concordances and Biblical Study." *Christian Century* 114, no. 32 (1997): 1034–37.

Kolko, Jed. "America's Most Diverse Neighborhoods and Metros." *Forbes*, November 11, 2012. http://www.forbes.com/sites/trulia/2012/11/13/finding-diversity-in-america/.

Kolodziejska, Marta, and A. L. P. Arat. "Religious Authority Online: Catholic Case Study in Poland." *Religion & Society in Central & Eastern Europe* 9, no. 1 (2016): 3–16.

Kozinets, Robert V. *Netnography: Ethnographic Research in the Age of the Internet*. Thousand Oaks, CA: Sage, 2010.

Krukowski, Damon. *Ways of Hearing*. Cambridge, MA: MIT Press, 2019.

Kruse, Kevin M. *One Nation under God: How Corporate America Invented Christian America*. New York: Basic Books, 2015.

Kucharsky, David. "Year of the Evangelical '76." *Christianity Today* 21, no. 2 (1976): 12–13.

Kyle, Richard G. *Evangelicalism: An Americanized Christianity*. New Brunswick, NJ: Transaction Publishers, 2006.

Kyle, Richard G. *Popular Evangelicalism in American Culture*. London: Routledge, 2017.

Labberton, Mark. "Opinion: Are Evangelicals Today More Devoted to Trump and the Republicans Than the Gospel?" *LA Times*, August 26, 2016. http://www.latimes.com/opinion/readersreact/la-ol-le-trump-evangelicals-white-supremacy-20170826-story.html.

Labberton, Mark. *Still Evangelical? Insiders Reconsider Political, Social and Theological Meaning*. Downers Grove, IL: InterVarsity Press, 2018.

Laughlin, Corrina. "'What God Gave to Us': Digital Habitus and the Shifting Social Imaginary of American Evangelicalism." PhD diss., University of Pennsylvania, 2018.

Leonard, Bill . "Getting Saved in America: Conversion Event in a Pluralistic Culture." *Review & Expositor* 82, no. 1 (Winter 1985): 111–27.

Levitin, Daniel J. *The Organized Mind: Thinking Straight in the Age of Information Overload*. Boston: Dutton, 2015.

LifeWay Research. "Protestant Pastors Name Graham Most Influential Living Preacher." *LifeWay Research*, February 2, 2010. https://lifewayresearch.com/2010/02/02/protestant-pastors-name-graham-most-influential-living-preacher/.

"Lifeway's New Bible Software Debuts among SBC Seminarians." Updated October 17, 2003. http://www.bpnews.net/16883/lifeways-new-bible-software-debuts-among-sbc-seminarians.

Lindemann, Sebastian. "Bye, by Burger!" Updated October 11, 2014. https://medium.com/startup-grind/bye-bye-burger-5bd963806015.

Lindsay, D. Michael. *Faith in the Halls of Power: How Evangelicals Joined the American Elite*. Oxford: Oxford University Press, 2007.

Liu, Ziming. *Paper to Digital: Documents in the Information Age*. Westport, CT: Libraries Unlimited, 2008.

Liu, Ziming. "Reading Behaviour in the Digital Environment: Changes in Reading Behaviour over the Past Ten Years." *Journal of Documentation* 61, no. 6 (2005): 700–12. http://dx.doi.org/10.1108/00220410510632040.

Logan, Sarah, and Rhona Johnston. "Investigating Gender Differences in Reading." *Educational Review* 62, no. 2 (June 16, 2010): 175–87.

"Logos Bible Software 3—Scholar's Library: Gold." *Bible Software Review*, updated June 2, 2007. http://web.archive.org/web/20111011084122/http://www.bsreview.org/index.php?modulo=Reviews&id=10.

"Logos Bible Software Rebrands as Faithlife, Acquires Beacon Ads." Logos, updated October 7, 2014. https://web.archive.org/web/20160304091759/https://www.logos.com/press/releases/Logos-Bible-Software-rebrands-as-Faithlife-acquires-Beacon-Ads.

Lövheim, Mia, and Heidi A. Campbell. "Considering Critical Methods and Theoretical Lenses in Digital Religion Studies." *New Media & Society* 19, no. 1 (2017): 5–14. https://doi.org/10.1177/1461444816649911.

Lubeck, Scott. "Scrolling through the Bible." *Texas Monthly*, December 1984, 206–8.

Luhrmann, Tanya *When God Talks Back: Understanding the American Evangelical Relationship with God*. New York: Vintage, 2012.

Lum, Casey Man Kong, ed. *Perspectives on Culture, Technology and Communication: The Media Ecology Tradition*. New York: Hampton Press Communication, 2005.

MacArthur, John. *Ashamed of the Gospel: When the Church Becomes Like the World*. 3rd ed. Wheaton, IL: Crossway Books, 2010.

MacKenzie, Donald, and Judy Wajcman, eds. *The Social Shaping of Technology*. Milton Keynes, England: Open University Press, 1985.

Maehre, Bob, and Kenneth R. Wade. "Computer Concordances." *Ministry*, November 1987. http://adventistdigitallibrary.org/islandora/object/adl:376011/datastream/PDF/view.

Malley, Brian. *How the Bible Works: An Anthropological Study of Evangelical Biblicism*. Walnut Creek, CA: AltaMira Press, 2004.

Malley, Brian. "Understanding the Bible's Influence." In *The Social Life of Scriptures: Cross-Cultural Perspectives on Biblicism*, edited by James Bielo, 194–204. New Brunswick, NJ: Rutgers University Press, 2009.

Mangen, Anne. "Hypertext Fiction Reading: Haptics and Immersion." *Journal of Research in Reading* 31, no. 4 (2008): 404–19. http://dx.doi.org/10.1111/j.1467-9817.2008.00380.x.

Mangen, Anne, P. Robinet, G. Olivier, and J. L. Velay. "Mystery Story Reading in Pocket Print Book and on Kindle: Possible Impact on Chronological Events Memory." Paper presented at a conference of the International Society for the Empirical Study of Literature and Media, Turin, Italy, July 21–25, 2014.

Mangen, Anne, Bente R. Walgermo, and Kolbjørn Brønnick. "Reading Linear Texts on Paper versus Computer Screen: Effects on Reading Comprehension." *International Journal of Educational Research* 58 (2013): 61–68.

Margolin, Sara J., Casey Driscoll, Michael J. Toland, and Jennifer Little Kegler. "E-readers, Computer Screens, or Paper: Does Reading Comprehension Change across Media Platforms?" *Applied Cognitive Psychology* 27, no. 4 (2013): 512–19. https://doi.org/10.1002/acp.2930.

Marsden, George M. *Fundamentalism and American Culture*. 2nd ed. Oxford: Oxford University Press, 2006.

Marsden, George M. *Reforming Fundamentalism: Fuller Seminary and the New Evangelicalism*. Grand Rapids, MI: Eerdmans, 1987.

Martínez, Jessica, and Gregory A. Smith. "How the Faithful Voted: A Preliminary 2016 Analysis." Pew Research, updated November 9, 2016. http://www.pewresearch.org/fact-tank/2016/11/09/how-the-faithful-voted-a-preliminary-2016-analysis/.

Marx, Leo, ed. *Does Technology Drive History? The Dilemma of Technological Determinism*. Cambridge, MA: MIT Press, 1994.

Mattox, Mickey L. "Martin Luther." In *Christian Theologies of Scripture: A Comparative Introduction*, edited by Justin Holcomb, 94–113. New York: New York University Press, 2006.

McBride, Jason. "In the '60s, Marshall Mcluhan Was Toronto's Most Famous Intellectual; Now, the World Has Finally Caught up with Him." *Toronto Life* (Toronto), July 6, 2011. https://torontolife.com/city/marshall-mcluhan-profile/.

McCuen, Gary E. *The Religious Right*. Hudson, WI: G.E. McCuen Publications, 1989.

McGavran, Donald A. *Understanding Church Growth*. Grand Rapids, MI: Eerdmans, 1970.

McGrath, Alister. *Evangelicalism and the Future of Christianity*. London: Hodder & Stoughton, 1993.

McKinley, Mike. *Church Planting Is for Wimps: How God Uses Messed-Up People to Plant Ordinary Churches That Do Extraordinary Things*. Wheaton, IL: Crossway Books, 2010.

McLane, Adam. "The Opportunities Smartphones Present for Your Church." *The Evangelicals* 4, no. 1 (Spring/Summer 2018): 14–17.

McLuhan, Marshall. *The Gutenberg Galaxy: The Making of Typographic Man*. Toronto: University of Toronto Press, 1962.

McLuhan, Marshall. *Understanding Media: The Extensions of Man*. New York: New American Library, 1964.

McLuhan, Marshall, and Quentin Fiore. *The Medium Is the Massage*. New York: Bantam, 1967.

McShane, Clay. "The Centrality of the Horse to the Nineteenth-Century American City." In *The Making of Urban America*, edited by Raymond Mohl, 105–30. New York: SR Publishers, 1997.

Meacham, Jon. "The Editor's Desk." *Newsweek*, November 12, 2006. http://www.newsweek.com/editors-desk-106637.

"Memos." *Courier-Journal* (Louisville, Kentucky), December 17, 1984. https://www.newspapers.com/newspage/109761475/.

"Millennials Stand Out for Their Technology Use, but Older Generations Also Embrace Digital Life." Pew Research Center, updated May 2, 2018. https://www.pewresearch.org/fact-tank/2018/05/02/millennials-stand-out-for-their-technology-use-but-older-generations-also-embrace-digital-life/.

Miller, Calvin. *The Vanishing Evangelical: Saving the Church from Its Own Success by Restoring What Really Matters*. Grand Rapids, MI: Baker Books, 2013.

"Mission: About Our Bible App." 2018, accessed September, 2018. https://www.ourbibleapp.com/mission/.

Mobile Fact Sheet. Pew Research Center, 2017, accessed March 22, 2018. http://www.pewinternet.org/fact-sheet/mobile/.
Mohler, Albert. "Confessional Evangelicalism." In *Four Views on the Spectrum of Evangelicalism*, edited by Collin Hansen, 68–96. Grand Rapids, MI: Zondervan, 2011.
Monsma, Stephen J. *Responsible Technology: A Christian Perspective*. Grand Rapids, MI: Eerdmans, 1986.
Montoya, Susan. "Churches Take 'Baptist' from Name." *AP News*, January 12, 1999. https://apnews.com/article/841aabf08eedd8c220aa7683cfa72f37.
More, Thomas. *The Yale Edition of the Complete Works of St. Thomas More*. Edited by Louis A. Schuster, Richard C. Marius, and James P. Lusardi. Vol. 8. New Haven, CT: Yale University Press, 1973.
Moreton, Bethany. *To Serve God and Wal-Mart: The Making of Christian Free Enterprise* Cambridge, MA: Harvard University Press, 2010.
Morgan, James, and Jeffrey K. Liker. *The Toyota Product Development System: Integrating People, Process, and Technology*. London: Taylor & Francis, 2006.
Mosemghvdlishvili, Lela, and Jeroen Jansz. "Negotiability of Technology and Its Limitations." *Information, Communication & Society* 16, no. 10 (2013): 1596–618. https://doi.org/10.1080/1369118X.2012.735252.
Moss, Candida R., and Joel S. Baden. *Bible Nation: The United States of Hobby Lobby*. Princeton, NJ: Princeton University Press, 2017.
Mouw, Richard J. *Talking with Mormons: An Invitation to Evangelicals*. Grand Rapids, MI: Eerdmans, 2012.
Mullins, Phil. "Imagining the Bible in Electronic Culture." *Religion & Education* 23, no. 1 (June 1, 1996): 38–45. https://doi.org/10.1080/15507394.1996.11000822.
Murashko, Alex. "'The Bible' Producers Announce Upcoming Youversion Companion App." *Christian Post*, February 8, 2013. http://www.christianpost.com/news/the-bible-producers-announce-upcoming-youversion-companion-app-89696/.
"My Take: How Technology Could Bring Down the Church." *CNN*, updated May 15, 2011. http://religion.blogs.cnn.com/2011/05/15/my-take-how-technology-could-bring-down-the-church/comment-page-8/.
National Association of Evangelicals. "Context." *Evangelicals*, 3, no. 3 (Winter 2017): 5.
Nel, Malan, and Eric Scholtz. "Calling, Is There Anything Special about It?" *Hervormde Teologiese Studies* 72, no. 4 (2016): 1–7. https://doi.org/10.4102/hts.v72i4.3183.
Nelson, Peter K. "Impractical Christianity: Faith Really Begins to Make a Difference When It Stops 'Working.'" *Christianity Today* 49, no. 9 (2005): 80–82.
"The New English Translation (NET) Bible Available in Print with 60,000 Translators' Notes." *Bible Gateway*, updated August 14, 2019. https://www.biblegateway.com/blog/2019/08/the-new-english-translation-net-bible-available-in-print-with-60000-translators-notes/.
Nicholaou, Nick B. "The Little Tools I Love." *Clergy Journal* 83, no. 2 (2006): 25.
Nicolet, Lindsay. "The Bible as an App." *Light* 4, no. 1 (2018): 16–18.
Niederle, Muriel, and Lise Vesterlund. "Do Women Shy Away from Competition? Do Men Compete Too Much?" *Quarterly Journal of Economics* 122, no. 3 (2007): 1067–101. https://doi.org/10.1162/qjec.122.3.1067. https://doi.org/10.1162/qjec.122.3.1067.
Noll, Mark A. *American Evangelical Christianity: An Introduction*. Oxford: Blackwell, 2001.
Noll, Mark A. "What Has Wheaton to Do with Jerusalem? Lessons from Evangelicals for the Reformed." *Reformed Journal* 32, no. 5 (1982): 8–15.
Noll, Mark A. "What Is 'Evangelical'?" In *The Oxford Handbook of Evangelical Theology*, edited by Gerald R. McDermott, 19–34. Oxford: Oxford University Press, 2010.

"The Non-Trump Evangelicals." *The Economist*, April 19, 2018. https://www.economist.com/united-states/2018/04/19/the-non-trump-evangelicals.

Norris, Pippa, and Ronald Inglehart. *Sacred and Secular: Religion and Politics Worldwide*. Cambridge: Cambridge University Press, 2011.

Noyes, J. M., and K. J. Garland. "Computer- vs. Paper-Based Tasks: Are They Equivalent?" *Ergonomics* 51, no. 9 (September 2008): 1352–75. https://doi.org/10.1080/00140130802170387.

O'Donnell, Matthew Brook. "Linguistic Fingerprints or Style by Numbers? The Use of Statistics in the Discussion of Authorship of New Testament Documents." In *Linguistics and the New Testament: Critical Junctures*, edited by Stanley E. Porter and D. A. Carson, 148–88. Sheffield: Sheffield Academic Press, 1999.

O'Leary, Stephen D. "Cyberspace as Sacred Space: Communicating Religion on Computer Networks." *Journal of American Academy of Religion* 64, no. 4 (1996): 781–808.

O'Leary, Stephen D., and Brenda Brasher. "The Unknown God of the Internet: Religious Communication from the Ancient Agora to the Virtual Forum." In *Philosophical Perspectives on Computer-Mediated Communication*, edited by Charles Ess, 233–67. Albany: State University of New York Press, 1996.

O'Neill, J. "Digital Annotation: Not There Yet." *Information Today* 27, no. 7 (July 2010): 1+.

Ockenga, Harold, and Leslie R. Marston. *Our Evangelical Faith*. Whitefish, MT: Literary Licensing, 1946.

"On 25th Anniversary, Bible Gateway Has Been Viewed More Than 14 Billion Times." *Bible Gateway*, updated August 1, 2018. https://www.biblegateway.com/blog/2018/08/on-25th-anniversary-bible-gateway-has-been-viewed-more-than-14-billion-times/.

"Online Bible Homepage." 2018, accessed February 24, 2017. http://web.archive.org/web/20080312151153/http://www.onlinebible.org/.

Ophira, Eyal, Clifford Nass, and Anthony D. Wagner. "Cognitive Control in Media Multitaskers." *Proceedings of the National Academy of Sciences* 106, no. 33 (2009): 15583–87.

Ovwigho, Pamela Caudill, and Arnold Cole. "Scriptural Engagement, Communication with God, and Moral Behavior among Children." *International Journal of Children's Spirituality* 15, no. 2 (2010): 101–13. https://doi.org/10.1080/1364436X.2010.497642.

Ovwigho, Pamela Caudill, and Arnold Cole. "Understanding the Bible Engagement Challenge: Scientific Evidence for the Power of 4." December 2009. https://web.archive.org/web/20161215103917/http://www.backtothebible.org/files/web/docs/cbe/Scientific_Evidence_for_the_Power_of_4.pdf.

Packer, J. I. *The Evangelical Anglican Identity Problem: An Analysis*. Latimer, England: Latimer House, 1978.

Packer, J. I., and Thomas C. Oden. *One Faith: The Evangelical Consensus*. Westmont, IL: IVP Books, 2004.

Page, Melvin Eugene, and Penny M. Sonnenburg. *Colonialism: An International, Social, Cultural, and Political Encyclopedia (3 Volumes)*. Santa Barbara, Denver, and Oxford: ABC-CLIO, 2003.

Parker, David. "Logos Bible Software Series X Scholar's Library Biblical Languages Supplement." *Evangelical Review of Theology* 27, no. 3 (2003): 272–75.

Patrick, Darrin. *Church Planter: The Man, the Message, the Mission*. Wheaton, IL: Crossway, 2010.

Payne, Jervis David. *Apostolic Church Planting: Birthing New Churches from New Believers*. Downers Grove, IL: InterVarsity Press, 2015.

Pear, Robert. "Trump Administration Set to Roll Back Birth Control Mandate." *New York Times*, October 5, 2017. https://www.nytimes.com/2017/10/05/us/politics/trump-birth-control.html.

Perry, Samuel L. *Addicted to Lust: Pornography in the Lives of Conservative Protestants*. New York: Oxford University Press, 2019.

Pew Research Center. "Four-in-Ten Americans Credit Technology with Improving Life Most in the Past 50 Years." Pew Research Center, 2017, accessed November 24, 2018. http://www.pewresearch.org/fact-tank/2017/10/12/four-in-ten-americans-credit-technology-with-improving-life-most-in-the-past-50-years/.

Pew Research Center. "Religious Landscape Study." 2014, accessed March 17, 2017. http://www.pewforum.org/religious-landscape-study/.

Pew Research Center. *Religious Landscape Study*. 2018. http://www.pewforum.org/religious-landscape-study/gender-composition/.

Pew Research Center. *Technology Device Ownership: 2015*. Pew Research Center, October 2015. http://www.pewinternet.org/2015/10/29/technology-device-ownership-2015.

Phillips, Peter M. *The Bible, Social Media and Digital Culture*. London: Routledge, 2019.

Piolat, Annie, Jean-Yves Roussey, and Olivier Thunin. "Effects of Screen Presentation on Text Reading and Revising." *International Journal of Human-Computer Studies* 47 (1996): 565–89.

Pitt, Richard N. *Divine Callings: Understanding the Call to Ministry in Black Pentecostalism*. New York: New York University Press, 2012.

Pollack, Andrew. "Putting Out the Word Electronically." *New York Times*, August 29, 1982. http://www.nytimes.com/1982/08/29/weekinreview/putting-out-the-word-electronically.html.

Porter, Tecoy M., Sr. *Faith to Innovate: 21st Century Tools & Strategies for Leadership Transformation*. Elk Grove, CA: Inner Treasure Press, 2015.

Price, James Matthew. "Undergraduate Perceptions of Vocational Calling into Missions and Ministry." *Missiology* 41, no. 1 (2013): 87–96. https://doi.org/10.1177/0091829612466997.

Purcell, Kevin. "Interview with Stephen Johnson, CEO of Olivetree Bible Software." *Church Tech Today*, September 26, 2012. http://churchtechtoday.com/2012/09/26/interview-stephen-johnson-ceo-olivetree-bible-software/.

"Rainbow Studies Homepage." 1997. https://web.archive.org/web/19970327090752/http://www.rainbowstudies.com/.

Rainie, Lee, and Barry Wellman. *Networked: The New Social Operating System*. Cambridge, MA: MIT Press, 2012.

Rakow, Katja. "The Bible in the Digital Age: Negotiating the Limits of 'Bibleness' of Different Bible Media." In *Christianity and the Limits of Materiality*, edited by Minna Opas and Anna Haapalainen, 101–21. London: Bloomsbury, 2016.

"Read the Bible Online!" 2000, accessed July 23, 2017. http://web.archive.org/web/20000229202117/http://www.bible.com/bible_read.html.

"The Real Origins of the Religious Right." *Politico*, updated May 27, 2014. https://www.politico.com/magazine/story/2014/05/religious-right-real-origins-107133?paginate=false.

Reed, Eric. "The Best of Today's Preachers." *Christianity Today Pastors*, February 6, 2002. https://www.christianitytoday.com/pastors/2002/february-online-only/cln20206.html.

Reilly, David, David Neumann, and Glenda Andrews. "Gender Differences in Reading and Writing Achievement: Evidence from the National Assessment of Educational Progress (NAEP)." *American Psychologist* 74, no. 4 (2019): 446–47. https//:doi.org/10.1037/amp0000356.

Renaud, Myriam. "Myths Debunked: Why Did White Evangelical Christians Vote for Trump?" University of Chicago Divinity School, The Martin Marty Center for the Public Understanding of Religion, January 19, 2017. https://divinity.uchicago.edu/sightings/myths-debunked-why-did-white-evangelical-christians-vote-trump.

Rheingold, Howard. *The Virtual Community: Homesteading on the Electronic Frontier*. New York: Harper Perennial, 1993.

Richtel, Matt. "Digital Devices Deprive Brain of Needed Downtime." *New York Times*, August 24, 2010. https://www.nytimes.com/2010/08/25/technology/25brain.html.

Ridgeway, Chris. "Scripture in Digital Context: Explorations in Media Ecology and Theology." 2009, accessed July 23, 2017. https://www.chrisridgeway.net/thesis.

Roberts, Bill. "Mart Green Partners with Wycliffe, Biblica, Bible Society on Digital Bible Library." *Charisma News*, December 14, 2012. http://www.charismanews.com/us/34940-mart-green-partners-with-wycliffe-biblica-bible-society-on-digital-bible-library.

Roberts, Colin H. *The Codex*. London: British Academy, 1954.

Roberts, Colin H., and T. C. Skeat. *The Birth of the Codex*. Repr. ed. Oxford: Oxford University Press, 1987.

Rocherolle, Narendra. "The Origin of the Retweet and Other Twitter Arcana." Medium, updated March 6, 2019. https://medium.com/@narendra/the-origin-of-the-retweet-and-other-twitter-arcana-5c53289d9a47.

Roof, Wade Clark. *Spiritual Marketplace: Baby Boomers and the Remaking of American Religion*. Princeton, NJ: Princeton University Press, 1999.

Rosman, Doreen M. *Evangelicals and Culture*. 2nd ed. Cambridge: James Clarke, 2012.

Ross, Bobby, Jr. "With 330 Million Downloads, Top Bible App Celebrates 10 Years." *Religion News Service*, July 26, 2018. https://religionnews.com/2018/07/26/with-330-million-downloads-top-bible-app-celebrates-10-years/.

Ross, W. Gordon. "Christianity in a Computer Age." *Religion in Life* 34, no. 4 (1965): 631–32.

Rowston, D. J. "The Most Neglected Book in the New Testament." *New Testament Studies* 21, no. 4 (1975): 554–63.

Ryle, J. C. *Evangelical Religion: What It Is, and What It Is Not*. London: Willam Hunt and Company, 1867.

Sax, David. *The Revenge of Analog: Real Things and Why They Matter*. New York: PublicAffairs, 2016.

Schaposnik, Laura P., and James Unwin. "The Phone Walkers: A Study of Human Dependence on Inactive Mobile Devices." *eprint arXiv* 1804.08753 (April 25, 2018): 389–414.

Scheifinger, Heinz. "The Significance of Non-participatory Digital Religion: The Saiva Siddhanta Church and the Development of a Global Hinduism." In *Digital Hinduism: Dharma and Discourse in the Age of New Media*, edited by Murali Balaji, 3–24. Lanham, MD: Lexington Books, 2017.

Scholz, Jan, Tobias Selge, Max Stille, and Johannes Zimmermann. "Listening Communities? Some Remarks on the Construction of Religious Authority in Islamic Podcasts." *Welt des Islams* 48, nos. 3–4 (2008): 457–509. https://doi.org/10.1163/157006008X364721.

Schreiner, Thomas R. *1, 2 Peter, Jude*. Nashville: Broadman & Holman Publishers, 2003.
"Scripture Engagement Defined." Taylor University Center for Scripture Engagement, updated August 31, 2011. http://tucse.taylor.edu/scripture-engagement-defined/.
"The Scripture Scanner Advertisement." *PC Mag*, Febuaray 21, 1984.
"Scriptures.Com Homepage." 1999, accessed August 15, 2017. http://web.archive.org/web/19991013082020/http://scriptures.com/.
Schultze, Quentin J. *Habits of the High-Tech Heart: Living Virtuously in the Information Age*. Grand Rapids, MI: Baker Books, 2002.
Schumacher, Eric H., Travis L. Seymour, Jennifer M. Glass, David E. Fencsik, Erick J. Lauber, David E. Kieras, and David E. Meyer. "Virtually Perfect Time Sharing in Dual-Task Performance: Uncorking the Central Cognitive Bottleneck." *Psychological Science* 12, no. 2 (March 1, 2001): 101–8. https://doi.org/10.1111/1467-9280.00318.
Schutz, Samuel R. "The Truncated Gospel in Modern Evangelicalism: A Critique and Beginning Reconstruction." *Evangelical Review of Theology* 33, no. 4 (2009): 292–305.
Schuurman, Derek C. *Shaping a Digital World: Faith, Culture and Computer Technology*. Downers Grove, IL: IVP Academic, 2013.
Seales, William Brent, Clifford Seth Parker, Michael Segal, Emanuel Tov, Pnina Shor, and Yosef Porath. "From Damage to Discovery Via Virtual Unwrapping: Reading the Scroll from En-Gedi." *Science Advances* 2, no. 9 (2016): e1601247. https://doi.org/10.1126/sciadv.1601247. http://advances.sciencemag.org/content/2/9/e1601247.abstract.
Searcy, Nelson, and Kerrick Thomas. *Launch: Starting a New Church from Scratch*. Rev, ed. Grand Rapids, MI: Baker Books, 2017.
"Shareware: Study Tools." Updated October 6, 1999. http://web.archive.org/web/19991006083257/http://www.christianshareware.net/StudyTools/.
Sharot, Stephen. "Beyond Christianity: A Critique of the Rational Choice Theory of Religion from a Weberian and Comparative Religions Perspective." *Sociology of Religion* 63, no. 4 (2002): 427–54. https://doi.org/10.2307/3712301.
Shashaani, Lily, and Ashmad Khalili. "Gender and Computers: Similarities and Differences in Iranian College Students' Attitudes toward Computers." *Computers & Education* 37, no. 3 (November 1, 2001): 363–75. https://doi.org/https://doi.org/10.1016/S0360-1315(01)00059-8.
Shelley, Bruce. *Evangelicalism in America*. Grand Rapids, MI: Eerdmans, 1967.
Shelley, Cameron. "The Nature of Simplicity in Apple Design." *Design Journal* 18, no. 3 (2015): 439–56.
Shellnutt, Kate. "2020's Most-Read Bible Verse: 'Do Not Fear.'" *Christianity Today*, updated December 3, 2020. https://www.christianitytoday.com/news/2020/december/most-popular-verse-youversion-app-bible-gateway-fear-covid.html.
Shellnutt, Kate. "After Mass Shootings, Americans Turn to Four Bible Verses Most." *Christianity Today*, updated October 2, 2017. http://www.christianitytoday.com/news/2017/october/after-mass-shootings-top-bible-verses-psalm-34-18-las-vegas.html.
Shellnut, Kate. "#Churchtoo: Andy Savage Resigns from Megachurch over Past Abuse." *Christianity Today*, March 18, 2018. https://www.christianitytoday.com/news/2018/march/andy-savage-resigns-abuse-megachurch-standing-ovation.html.
Shellnut, Kate, and Sarah Eekhoff Zylstra. "Who's Who of Trump's 'Tremendous' Faith Advisers." *Christianity Today*, June 22, 2016. https://www.christianitytoday.com/ct/2016/june-web-only/whos-who-of-trumps-tremendous-faith-advisors.html.
Sherman, Amy L. *Kingdom Calling: Vocational Stewardship for the Common Good*. Downers Grove, IL: IVP Books, 2011.

Shontell, Alyson. "With 100 Million Downloads, YouVersion Bible Is a Massive App That No VC Can Touch." *Business Insider*, updated July 29, 2013. https://www.businessinsider.com/youversion-bible-app-has-100-million-downloads-2013-7.

Siker, Jeffrey. *Liquid Scripture: The Bible in a Digital World*. Minneapolis: Fortress Press, 2017.

Skeat, T. C. "The Origin of the Christian Codex." *Zeitschrift für Papyrologie und Epigraphik* 102 (1994): 263–68.

Smietana, Bob. "Bible App Offers Portability for the Faithful." *USA Today*, August 13, 2013. https://www.usatoday.com/story/tech/2013/08/11/youversion-bible-app/2640719/.

Smith, Christian. *The Secular Revolution: Power, Interests, and Conflict in the Secularization of American Public Life*. Berkeley: University of California Press, 2003.

Smith, Christian, and Melinda Lundquist Denton. *Soul Searching: The Religious and Spiritual Lives of American Teenagers*. Oxford: Oxford University Press, 2005.

Smith, Christian, and Michael O. Emerson. *American Evangelicalism: Embattled and Thriving*. Chicago: University of Chicago Press, 1998.

Smith, Gary Scott. *Faith and the Presidency: From George Washington to George W. Bush*. Oxford: Oxford University Press, 2006.

Smith, Gordon T. *Courage and Calling: Embracing Your God-Given Potential*. Rev. ed. Downers Grove, IL: IVP Books, 2011.

Solomon, Brian. "Meet David Green: Hobby Lobby's Biblical Billionaire." *Forbes*, September 18, 2012. https://www.forbes.com/sites/briansolomon/2012/09/18/david-green-the-biblical-billionaire-backing-the-evangelical-movement/.

Stackhouse, John G., Jr. "Evangelicals and the Bible Yesterday, Today, and Tomorrow." In *New Paradigms for Bible Study: The Bible in the Third Millennium*, edited by Robert M. Fowler, Edith Blumhofer, and Fernando F. Segovia, 185–208. New York: Continuum, 2004.

Stetzer, Ed. "Defining Evangelicals in Research." *The Evangelicals* 3, no. 3 (2017–18): 12–13.

Stetzer, Ed. "Evangelism Never Changes, but the Methods Do." *Christianity Today*, February 6, 2015. https://www.christianitytoday.com/edstetzer/2015/february/evangelism-never-changes-but-never-stays-same.html.

Stetzer, Ed. "Fellow Evangelicals: Stop Falling for Trump's Anti-immigrant Rhetoric." *Vox*, November 2, 2018. https://www.vox.com/policy-and-politics/2018/11/6/18066116/trump-caravan-evangelical-voters.

Stewart, Kenneth J. "Did Evangelicalism Predate the Eighteenth Century? An Examination of David Bebbington's Thesis." *Evangelical Quarterly* 77, no. 2 (2005): 135–53.

Stott, John R. W. *Evangelical Truth: A Personal Plea for Unity, Integrity and Faithfulness*. Carlisle, PA: Langham Global Library, 2013.

Strate, Lance. *After Television in Trees: Echoes and Reflections on Media Ecology as a Field of Study*. New York: Hampton Press, 2006.

Strate, Lance. "President's Message: Understanding MEA." *In Medias Res* 1, no. 1 (1999): 1.

Subrahmanyam, Kaveri, Minas Michikyan, Christine Clemmons, Rogelio Carrillo, Yalda T. Uhls, and Patricia M. Greenfield. "Learning from Paper, Learning from Screens." *International Journal of Cyber Behavior, Psychology and Learning* 3, no. 4 (2013): 1–27. https://doi.org/10.4018/ijcbpl.2013100101.

Summers, Kate. "Adult Reading Habits and Preferences in Relation to Gender Differences." *Reference and User Services Quarterly* 52, no. 3 (2013): 243–49.

Sun, Szu-Yuan, Chich-Jen Shieh, and Kai-Ping Huang. "A Research on Comprehension Differences between Print and Screen Reading." *South African Journal of Economic and Management Sciences* 16 (2013): 87–101. http://www.scielo.org.za/scielo.php?script=sci_arttext&pid=S2222-34362013000500010&nrm=iso.

Surratt, Geoff, Greg Ligon, and Warren Bird. *The Multi-site Church Revolution: Being One Church—in Many Locations*. Grand Rapids, MI: Zondervan, 2006.

Sutton, Matthew Avery. *Jerry Falwell and the Rise of the Religious Right: A Brief History with Documents*. Boston: Bedford / St. Martin's, 2013.

Tanner, Kathryn. *Christianity and the New Spirit of Capitalism*. New Haven, CT: Yale University Press, 2019.

Tate, Allan. *Essays of Four Decades*. Wilmington, DE: ISI Books, 1999.

"This Is the Most Popular Verse in 2 Billion Pageviews during 2018 on Bible Gateway." *Bible Gateway*, updated December 10, 2018. https://www.biblegateway.com/blog/2018/12/this-is-the-most-popular-verse-in-2-billion-pageviews-during-2018-on-bible-gateway/.

Thomas, Jeremy N., and Andrew L. Whitehead. "Evangelical Elites' Anti-homosexuality Narratives as a Resistance Strategy against Attribution Effects." *Journal for the Scientific Study of Religion* 54, no. 2 (2015): 345–62. https://doi.org/10.1111/jssr.12188.

Tomkins, Stephen. *John Wesley: A Biography*. Grand Rapids, MI: Eerdmans, 2003.

Torma, Ryan, and Paul Emerson Teusner. "iReligion." *Studies in World Christianity* 17, no. 2 (2011): 137–55. https://doi.org/10.3366/swc.2011.0017.

Torry, Malcolm. *Managing God's Business: Religious and Faith-Based Organizations and Their Management*. New York: Routledge, 2016.

Trzebiatowska, Marta, and Steve Bruce. *Why Are Women More Religious Than Men?* Oxford: Oxford University Press, 2012.

Tsuria, Ruth, Aya Yadlin-Segal, Alessandra Vitullo, and Heidi A. Campbell. "Approaches to Digital Methods in Studies of Digital Religion." *Communication Review* 20, no. 2 (2017): 73–97. https://doi.org/10.1080/10714421.2017.1304137.

Tveit, Åse, and Anne Mangen. "A Joker in the Class: Teenage Readers' Attitudes and Preferences to Reading on Different Devices." *Library & Information Science Research* 35, nos. 3–4 (2014): 179–84.

Uncapher, Melina R., and Anthony D. Wagner. "Minds and Brains of Media Multitaskers: Current Findings and Future Directions." *Proceedings of the National Academy of Sciences* 115, no. 40 (2018): 9889. https://doi.org/10.1073/pnas.1611612115. http://www.pnas.org/content/115/40/9889.abstract.

US Census Bureau. *Home Computers and Internet Use in the United States: August 2000*. United States Census Bureau (September 2001). https://www.census.gov/prod/2001pubs/p23-207.pdf.

Vanhoozer, Kevin J. *The Drama of Doctrine: A Canonical Linguistic Approach to Christian Doctrine*. Louisville: Westminster John Knox Press, 2005.

Vaniman, Eric. "Select Scrolling Options in Your Logos Bible App." *Logos Talk*, March 19, 2015. https://web.archive.org/web/20190413225054/https://blog.logos.com/2015/03/select-scrolling-options-in-your-logos-bible-app/.

von Hippel, Eric. *The Sources of Innovation*. Oxford: Oxford University Press, 1998.

Wallace, Daniel B. "Innovations in Text and Translation of the Net Bible, New Testament." *Bible Translator* 52, no. 3 (2001): 335–49.

Wallace, Daniel B. "An Open Letter Regarding the Net Bible, New Testament." *Notes on Translation* 14, no. 3 (2000): 1–8.

Ward, W. Reginald. *Early Evangelicalism: A Global Intellectual History, 1670–1789*. Cambridge: Cambridge University Press, 2006.

Warner, R. Stephen. "Theoretical Barries to the Understanding of Evangelical Christianity." *Sociological Analysis* 40, no. 1 (1977): 1–9.

Wästlund, Erik, Henrik Reinikka, Torsten Norlander, and Trevor Archer. "Effects of VDT and Paper Presentation on Consumption and Production of Information: Psychological and Physiological Factors." *Computers in Human Behavior* 21, no. 2 (2005): 377–94. https://doi.org/10.1016/j.chb.2004.02.007.

Wayne, F. Stanford. "An Instrument to Measure Adherence to the Protestant Ethic and Contemporary Work Values." *Journal of Business Ethics* 8 (1989): 793–804.

Weber, Jeremy. "Why Zondervan Bought Biblegateway.Com." *Christianity Today*, November 6, 2008. http://www.christianitytoday.com/ct/2008/novemberweb-only/145-41.0.html.

Weber, Timothy. "Premillenialism and the Branches of Evangelicalism." In *The Variety of American Evangelicalism*, edited by Donald W. Dayton and Robert K. Johnston, 12–14. Knoxville: University of Tennessee Press, 2001.

Wegner, Daniel M., and Adrian F. Ward. "The Internet Has Become the External Hard Drive for Our Memories." *Scientific American* 309, no. 6: 58–61, accessed December 1, 2013. https://www.scientificamerican.com/article/the-internet-has-become-the-external-hard-drive-for-our-memories/.

"Westminster Confession of Faith." 1646, accessed May 24, 2022. http://www.ccel.org/ccel/anonymous/westminster3.i.i.html.

"What Is the Gramcord Institute?" N.d., accessed October 3, 2016. http://www.gramcord.org/whatis.htm.

"What's Good about Having the Bible on Mobile Devices? What's Dangerous?" Brookside Institute, updated September 3, 2015. https://www.thebrooksideinstitute.net/blog/whats-good-about-having-the-bible-on-mobile-devices-whats-dangerous.

Wigner, Dann. *A Sociology of Mystic Practices: Use and Adaptation in the Emergent Church*. Eugene, OR: Pickwick Publications, 2018.

Wilcox, Clyde, and Carin Robinson. *Onward Christian Soldiers? The Religious Right in American Politics*. 4th ed. Boulder, CO: Westview Press, 2011.

William, Gregg. "Review: The Apple Macintosh Computer." *Byte* 9, no. 2 (1984): 30–54, accessed August 22, 2019. https://archive.org/details/byte-magazine-1984-02/page/n53/mode/2up.

Williams, Robin, and David Edge. "The Social Shaping of Technology." *Research Policy* 25 (1996): 865–99.

Winner, Langdon. "Upon Opening the Black Box and Finding It Empty: Social Constructivism and the Philosophy of Technology." *Science, Technology, and Human Values* 18, no. 3 (1993): 362–78.

Witherington, Ben, III. *Work: A Kingdom Perspective on Labor*. Grand Rapids, MI: Eerdmans, 2011.

Wolf, Maryanne, and Catherine J. Stoodley. *Proust and the Squid: The Story and Science of the Reading Brain*. New York: HarperCollins, 2007.

Wolf, Maryanne, and Catherine J. Stoodley. *Reader, Come Home: The Reading Brain in a Digital World*. New York: Harper, 2018.

Wood, Molly. "Narcissist's Dream: Selfie-Friendly Phone." *New York Times*, February 6, 2014. https://www.nytimes.com/2014/02/06/technology/personaltech/making-the-case-for-a-more-selfie-friendly-smartphone.html.

"Wordsearch and Epiphany Software Join Forces." WORDSearch, updated July 14, 2003. http://web.archive.org/web/20110718113235/http://www.wordsearchbible.com/about/pressreleases/pr-epiphany-wordsearch-merger.php.
"Wordsearch Bible Homepage." 2017, accessed January 23, 2017. https://www.wordsearchbible.com.
"Wordsearch Bible Is Transitioning to Logos." Updated September 21, 2020. https://blog.faithlife.com/blog/2020/09/wordsearch-bible-is-transitioning-to-logos/.
Worthen, Molly. *Apostles of Reason: The Crisis of Authority in American Evangelicalism*. Oxford: Oxford University Press, 2016.
Wozniak, Kenneth W. M. "Evangelicals and the Ethics of Information Technology." *Journal of the Evangelical Theological Society* 28, no. 3 (1985): 335–42.
Wright, Bradley. "Conspicuous Consumption and Your iPhone." 2008, accessed March 17, 2017. http://www.everydaysociologyblog.com/2008/09/conspicuous-con.html.
Wuthnow, Robert. *Boundless Faith: The Global Outreach of American Churches*. Berkeley: University of California Press, 2009.
Wuthnow, Robert. *Rough Country: How Texas Became America's Most Powerful Bible-Belt State*. Princeton, NJ: Princeton University Press, 2014.
Wuthnow, Robert. *Sharing the Journey: Support Groups and America's New Quest for Community*. London: Free Press, 1996.
Wuthnow, Robert. "Taking Talk Seriously: Religious Discourse as Social Practice." *Journal for the Scientific Study of Religion* 50, no. 1 (2011): 1–21.
Young, Glenn. "Reading and Praying Online: The Continuity of Religion Online and Online Religion in Internet Christianity." In *Religion Online: Finding Faith on the Internet*, edited by Douglas Cowan and Lorne Dawson, 93–106. New York: Routledge, 2004.
YouVersion. "The Bible App for Kids Is Here!" April 12, 2013. https://blog.youversion.com/2013/11/the-bible-app-for-kids-is-here/.
YouVersion. "Bible.Com Joins the YouVersion Family!" YouVersion Blog, updated August 30, 2012. http://blog.youversion.com/2012/08/bible-com/.
YouVersion. "Easter Bible Engagement around the World." Updated April 6, 2015. http://blog.youversion.com/2015/04/easter-bible-engagement-around-the-world/.
YouVersion. "Infographics." 2014. http://blog.youversion.com/infographics/.
YouVersion. "Why Friendships in Bible App 5 Are Different." YouVersion, updated April 8, 2014, accessed July 23, 2017. http://blog.youversion.com/2014/04/why-friendships-in-bible-app-5-are-different/.
YouVersion. "YouVersion Expects Record-Breaking Bible Plan Completions This Easter." Updated April 8, 2019. https://www.youversion.com/press/youversion-expects-record-breaking-bible-plan-completions-this-easter/.
YouVersion. "YouVersion Homepage." 2017, accessed July 23, 2017. http://web.archive.org/web/20070705040140/http://www.youversion.com/.
Zylstra, Sarah Eekhoff. "What the Latest Bible Research Reveals about Millennials." *Christianity Today*, May 16, 2016. https://www.christianitytoday.com/news/2016/may/what-latest-bible-research-reveals-about-millennials.html.

Index

For the benefit of digital users, indexed terms that span two pages (e.g., 52–53) may, on occasion, appear on only one of those pages.

Accordance, 8, 25, 65, 80, 82–83
advertising, 47, 63–64, 77
Android, 16, 68, 79, 80
App Store, 25, 81, 110, 115
audio, 6, 7–8, 10–11, 16, 58, 80–81, 82–83, 124–25, 132, 132*f*, 133, 135, 147–48, 149–50, 151, 158–59, 173–74, 180, 185–86, 187
authority, 4, 13–14, 15, 18–19, 20, 21–24, 33, 36–37, 41, 80–81, 92–93, 105, 141–42, 178

Bible Gateway, 8, 9, 25–26, 71–73, 75–76, 77, 79, 82–83, 86, 88, 89, 91, 92–93, 94, 95–96, 97, 99, 100–1, 102, 103, 104, 107, 109, 111–12, 114, 117–18, 119, 136–37, 144, 182–83
Bitcoin, 43–44

Campbell, Heidi, 4, 18–20, 22, 23–24, 28–29, 86, 105, 140
children, 64–65, 92, 124–25, 126, 130, 134–35, 150–51, 173, 175, 176, 183–84
codex, 1–3, 7, 60–61, 181, 183–84
community, vii–viii, 22, 23–24, 26, 28–29, 38, 39, 41, 51–52, 92–93, 108, 115, 125–26, 182–83, 187
computer, vii, 1, 2–3, 6, 8, 10, 13–14, 16–17, 19, 20, 31, 41–42, 55–56, 58, 59, 60–61, 62–63, 64, 65, 67–68, 70, 72, 74, 78, 86, 87, 88–89, 124–25, 128, 130, 131–33, 134, 134*f*, 136, 142, 145, 147–48, 152, 153, 155–56, 157, 161, 174–75, 181, 182, 183–84

Dallas Theological Seminary (DTS), 66–67, 75–76, 84*f*
desktop, viii–3, 7–8, 10, 16–17, 25–26, 58, 61, 69– 70–, 72, 73, 82–83, 86, 103, 105–6, 129–30, 131–33, 145, 153, 161–62, 181–82, 183–84
devotional, 6–7, 9, 64–65, 102–3, 121–22, 124–25, 132, 133, 145–46, 149–50, 152, 154, 156, 183–84

English Standard Version (ESV), 10–11, 25, 82–83, 139
entrepreneurialism, 39–40, 45–46, 47, 48, 50, 65–66, 67–68, 76–77
  entrepreneur, 3, 45–48, 68–69, 81, 83, 92
  entrepreneurial, 5–6, 33, 41, 43–45, 46, 47–48, 62, 66, 68–69, 70–71, 74–75, 77, 179, 186–87
exegesis, exegetical, 8, 60–61, 65

Faithlife, 25, 51–52, 66, 68–69, 73, 93, 99–100, 101–2, 103, 105, 107, 108, 110, 111, 112, 114–15, 117, 120–21, 141, 182–83. *See also* Logos Bible Software
female, 82–83, 165, 169–73, 175

Greek, viii, 2, 30, 60–61, 67, 140
Gruenewald, Bobby, 81–82, 89

hamburger (menu), 115–16
hermeneutics, 7, 11, 186–87
Homogeneous Unit Principle (HUP), 46–47
Hopeful Entrepreneurial Pragmatism (HEP), 5–6, 24–25, 29, 42, 48, 56–57, 58, 60–61, 68–69, 71, 83–85, 86–87, 95, 107–8, 117, 122, 124, 147, 151, 154, 170, 171, 181–82, 184–85
Hutchings, Tim, 4, 23, 81–82, 100, 131–33, 132*f*, 173–74

identity, vii–viii, 1–2, 8–9, 19, 23–24, 31, 36, 37, 39, 45–46, 76–77, 81, 89–90, 99, 120–21, 125, 142, 144, 147, 180, 182, 186–87
interface, 15, 17–18, 83, 92, 103, 115–16, 161–62
interpretation, of the Bible, 20–21, 30–31, 37–38, 45, 99, 138, 166–67, 184, 185–86
iPhone, 8, 16, 68, 77–78, 79, 80, 81, 150–51, 152

laptop, 6, 10, 131–33
Life.Church, 25, 98, 117–18
Logos Bible Software, 8, 16, 25–26, 58, 65, 66–67, 72–73, 84*f*, 86, 88, 89
Logos, 8, 16, 25–26, 58, 65–67, 68–70, 72–74, 79–80, 82–83, 84*f*, 86, 88, 89, 93, 99, 100–2, 103, 104, 106, 108, 111, 112, 115, 117–19, 120–21, 133, 182–83

male, 32, 165, 168–66, 173, 174–75
marketing, 74, 103
materiality of the Bible, 78–79, 128, 129–30, 144
memorize, 80, 187
  memorization, 78–79, 80, 82–83, 132
mission, missions, Christian, 4–6, 31, 40, 41, 46–47, 48, 49–52, 55–56, 62, 72, 87–88, 91, 92, 96–97, 101, 116, 118–19, 120, 123, 156, 182
multimedia, 7, 132*f*, 139, 146, 148, 151, 187–88

Navigators, The, 65, 72
Nearest Available Bible (NAB), 6–7, 153, 154, 183–84
New American Standard Bible (NASB), 72–73, 75–76
New English Translation (NET), 8, 74, 75–77, 79, 80–81, 84*f*
New International Version (NIV), 25, 71, 72–73, 80, 139
notification, 5–6, 170, 175–77, 178–80, 184–85

open source, 21

pastor, vii–viii, 8, 26, 35–36, 42–43, 46, 47, 49, 54–55, 61, 63–65, 68, 69, 71, 73–74, 77, 78, 83–85, 90, 98, 99, 100–1, 102, 110–11, 123, 129–30, 182–83
PC Study Bible, 8, 22–23, 25, 65, 68
persuasive computing, 4, 171–72, 186–87
phone, smartphone, 6–8, 10–11, 14, 16, 17, 23, 27, 31, 39, 40–42, 55–57, 78–79, 80, 81–82, 86, 93, 110, 124–25, 128, 130, 131–33, 134, 134*f*, 135–36, 137–39, 142, 144, 145–46, 147–48, 149, 151–52, 153, 154–56, 157, 158–59, 160, 161, 162, 164–65, 166, 167, 168, 169–72, 174, 175–77, 178–80, 183–85, 187–88
podcast, 7–8, 10, 22, 42
politics, 4–5, 62, 123
  political, 13–14, 16–17, 18–19, 33–34, 35–36, 39, 45, 50, 51, 52, 53–54, 55–56
  politician, 45–46
portable, portability, 77–79, 148, 149, 153, 160
pragmatism, 4–6, 31, 35–36, 39–40, 48–53, 54–56, 61–62, 63, 64, 65, 68–69, 70–71, 97, 101–2, 115–16, 124–25, 126, 149–50, 152–53, 154, 181–82, 186–87
Pritchett, Bob, 66–67, 68–69, 89, 117
Protestant, 2, 22–23, 30–31, 33–34, 35, 38, 44, 45, 52–53, 56, 63, 157–58, 184

reading
  Bible reading, 1–3, 6–7, 8, 10–11, 15, 21–22, 25, 27, 36–37, 38–39, 42–43, 56–57, 70, 72, 78–79, 80–82, 86–87, 92, 94, 95, 96, 98, 99, 103, 108, 112, 113, 117, 119, 121–22, 124–25, 126–28, 129, 133–35, 139, 146, 147–48, 149–51, 153–56, 157–58, 160, 173–74, 175–77, 178–79, 182–86
  daily reading, 4, 27, 81–82, 83–85, 98, 102–3, 126–27, 133, 149, 150–51, 152, 153–54, 156, 160, 163–64, 168, 169, 170, 172, 174, 175, 178, 179–80, 184–86
  general reading, 5–6, 17, 94, 139, 161–64, 173–75, 178–80, 185–86

interpretive reading, 4, 5–6, 8–9, 20–21, 92, 141, 166–68, 182–83
reading comprehension, 6–7, 26–27, 125, 126, 160, 161–65, 167, 169, 170, 172, 173–74, 175, 179–80, 184–86
reading plans, 6, 16, 27, 37, 80, 81–82, 94, 103, 113, 121–22, 152, 160, 168, 169, 170–71, 172, 173–74, 175, 178, 180, 184–85
Religious Social Shaping of Technology (RSST), 23–25, 27, 28–29, 39, 86–87, 105, 126
ritual, 22, 44, 180
Roman Catholic, Catholicism, 15, 22–23, 30–31, 38, 44, 50–51, 61, 63, 105–6

screen, 1, 2–3, 4, 6–7, 16, 17, 59, 62–63, 77, 86–87, 92, 111–12, 115, 133, 135, 136, 137–38, 139, 143, 144, 146, 147–48, 153–54, 156, 157–59, 160, 161–64, 165, 166–68, 173–75, 177, 178–80, 183–86, 187
scroll (noun), 1–3, 60, 181, 183–84
scroll (verb), 16, 26–27, 89–90, 112, 114, 161–62
search, 1, 6–7, 9, 10, 25
seminary, vii–viii, 35, 36, 41, 66–67, 75–76, 81–82, 84*f*, 100
Social Construction of Technology (SCOT), 14–19, 20–21, 27, 28–29, 77, 86–87, 105–6, 109, 114–15
social media, 7–8, 9, 16–17, 23, 39–40, 42, 43–44, 113, 114, 133, 146, 147, 187
Social Shaping of Technology (SST), 12, 13–15, 17–19, 23, 76–77, 110, 111–12, 181–82
Storch, Terry, 89
study the Bible, 1–2, 4, 6, 7–8, 25, 31–32, 38–39, 46–47, 62–63, 65, 67, 68, 75, 76, 78–79, 80, 81, 92–93, 95, 99, 100–1, 102–3, 107, 111–12, 115, 121, 124–25, 128, 130, 132, 135–36, 138, 139, 140, 141–42, 144, 145, 146, 152, 156, 160, 162, 167–68, 172, 176–78, 181, 183–84

tablet, 6, 10, 17, 125, 128, 131–34, 134*f*, 147–48, 152, 162–63
transformation (spiritual), 6, 20, 45, 95–96, 98–99, 104, 121
translations, 2, 8, 21–22, 43–44, 51–52, 66–67, 70–71, 72–73, 74, 75–77, 80–81, 82–83, 110, 138–39, 140, 153, 187
Trump, Donald, 28, 35–36, 52–53, 54–56, 92–93

verse numbers, 2, 9, 10–11, 59, 80–82, 87, 102–3, 113–14, 130, 136–38, 139, 143, 146–47, 166–67, 171–72, 178

women, 10, 14, 21–22, 51, 86, 125–26, 169–70, 172–75
worship, 5, 18–19, 22–23, 38, 44, 78–79, 81, 124–25, 126, 134–35, 183–84

YouVersion, 4–5, 7–8, 9, 11, 16–17, 25–26, 27, 43–44, 51–52, 58, 71, 73–75, 77–83, 86, 89, 98–102, 104, 105, 106, 108, 112–14, 115–16, 117–19, 121–22, 123, 145, 146–47, 150, 168, 179–80, 182–83, 184–85

Zondervan, 25, 62, 66, 69–70, 72–73, 75–76, 79, 84*f*, 109